Agricultural Benefits of Postharvest Banana Plants

Authored by

Dibakar Chandra Deka
Madhabdev University,
Narayanpur, Assam,
India

&

Satya Ranjan Neog
Dhakuakhana College,
Lakhimpur, Assam,
India

Agricultural Benefits of Postharvest Banana Plants

Authors: Dibakar Chandra Deka and Satya Ranjan Neog

ISBN (Online): 978-981-18-0163-1

ISBN (Print): 978-981-18-0161-7

ISBN (Paperback): 978-981-18-0162-4

need for a court order if at any point you breach any terms of this License Agreement. In no event will any delay or failure by Bentham Science Publishers in enforcing your compliance with this License Agreement constitute a waiver of any of its rights.

3. You acknowledge that you have read this License Agreement, and agree to be bound by its terms and conditions. To the extent that any other terms and conditions presented on any website of Bentham Science Publishers conflict with, or are inconsistent with, the terms and conditions set out in this License Agreement, you acknowledge that the terms and conditions set out in this License Agreement shall prevail.

Bentham Science Publishers Pte. Ltd.
80 Robinson Road #02-00
Singapore 068898
Singapore
Email: subscriptions@benthamscience.net

BENTHAM SCIENCE

CONTENTS

PREFACE

This book reports a simple method of recycling banana farm bio-waste, thus helping farmers to make wealth out of waste.

Potassium is a major plant nutrient, and recycling it between plants and soil serves the best interest of both. Banana plant absorbs huge amount of potassium from soil and distributes between the trunk (pseudo-stem) and the fruits. Banana plants give fruits only once, and volume of pseudo-stem generated is five to ten times of fruits. Naturally, banana farming generates a huge quantity of biomasses and leads to severe depletion of soil potassium. This book reports how part of the depleted potassium can be restored to soil.

Banana is a major crop in at least 135 countries world over, and more than 150 million MT banana fruits are produced every year. This much of banana production is associated with 750 to 1500 million MT of bio-waste, and this much bio-waste is equivalent to 2.2213 to 4.4427 billion MT of muriate of potash (MOP). We are reporting to show how to use banana plant pseudo-stem in lieu of MOP to grow five different crops on experimental basis. Undoubtedly, our experiments may be extended to cover many other crops. The use of pseudo-stem juice as the substitute for potash not only restores soil potassium but also enhances crop yields minimum 10% up to about 60%.

The book consists of eleven chapters. The chapters include analysis of banana plant pseudo-stem juice and fibers. Details of farming procedures and crop yield analysis along with colored pictures are provided. Prospective uses of pseudo-stem fibers are also discussed. Further scope of research and development is discussed in the last chapter. A glossary of important terminologies and abbreviations is also provided for the convenience of the readers.

While conducting research, scientists should keep in mind the service to the society and must take utmost care to preserve the virginity of the environment. The use of banana plant pseudo-stem to grow other crops would serve both these dual purposes. It would bring additional value to banana farming, thus helping farmers in improving their economic conditions ; at the same time, it would protect the soil environment from harmful effects of chemical fertilizers. I wish that the objective of the book would be inspiring for others to take up works with similar spirits.

CONSENT FOR PUBLICATION

Not applicable.

CONFLICT OF INTEREST

The author declares no conflict of interest, financial or otherwise.

ACKNOWLEDGEMENTS

Declared none.

Dibakar Chandra Deka
Vice-Chancellor,
Madhabdev University
Narayanpur, Assam
India

&

Satya Ranjan Neog
Associate Professor of Chemistry,
Dhakuakhana College
Lakhimpur, Assam
India

CHAPTER 1

Introduction

Abstract: A brief introduction to banana plant and its different morphological parts has been presented. Traditional and reported uses of different morphological parts have been discussed. Post-harvest banana plant is of no use or little use. Keeping in mind the prospective uses of banana plant in lieu of potash in agriculture, reported non-renewable sources of potash of mineral origin have been discussed. Banana plant is a rich source of potassium chloride and potassium carbonate. The importance of these two chemicals and their reported sources and uses, have been discussed. A brief survey on banana producing countries across the globe, global majors of potash exporters and consumers is also presented. Towards the end, an outline of the book chapters can be seen.

Keywords: Banana plant, *Kolakhar*, Potash, Potassium carbonate, Pseudo-stem, Uses of morphological parts.

1. A BRIEF INTRODUCTION OF BANANA PLANT

The banana plant is a large herbaceous flowering plant. The size and height of the plant depend on the variety and growing conditions (Figs. **1** and **2**). The tall variety such as 'Gros Michel' may grow up to a height of 7 m (23 ft) and the height of 'Dwarf Cavendish' may be limited to around 2 m (7 ft). Most of the other varieties stand at around 5 m (16 ft) tall [1, 2]. The plant grows from a fleshy rhizome, which is referred to as 'corm'. The corm remains close to ground and the plant above ground appears like the trunk of a tree, but it is actually a "false stem" or pseudo-stem. The leaves of banana plants consist of a petiole and a lamina. The base of the petiole widens to form a sheath. Leaf-sheaths are spirally arranged and tightly packed to make the pseudo-stem of cylindrical shape (Figs. **3** and **4**). Leaves may grow to about 3 m (9 ft) long and 60 cm (2.0 ft) wide.

When matured, the corm of the banana plant stops producing new leaves. Instead, a flower spike or inflorescence is developed, and the immature inflorescence is pushed up by a growing stem along the centre of the pseudo-stem. The inflorescence eventually emerges at the top. Each plant normally produces a single inflorescence, which is often referred to as the 'banana heart'. The inflorescence contains rows of flowers with a bract between two rows. The bracts are sometimes incorrectly called petals. The rows of female flowers appear first, followed by the rows of male flowers. Female flowers develop into fruits to form a large hanging cluster consisting of multiple tiers. Each tier (called a 'hand') consists of up to 20 fruits. A hanging cluster comprising of several tiers may weigh up to over 50 kilograms. In cultivated varieties, the seeds virtually do not exist. Their remnants as tiny black specks are often visible in the interior of the fruits. Bananas display slight radioactivity because of the natural presence of the isotope potassium-40 in trace amount along with the bulk potassium [3, 4].

2. CLASSIFICATION OF BANANA PLANT

Banana plant belongs to kingdom *plantae*. Carl Linnaeus [5] classified banana plant as follows:

Kingdom: Plantae

Division/Clade: Angiosperms

Order: Zingiberales

Family: Musaceae

Genus: *Musa*

Classification of banana plant has always been a problem for taxonomists because of the existence of large number of hybrids arising from hybridization among the species of genus *Musa* [2].

Based on the uses, bananas were originally classified by Linnaeus into two species - *Musa sapientum* for dessert bananas and *Musa paradisiaca* for plantains [6]. This simplistic classification was not adequate to address the large number of cultivars that subsequently emerged [7]. Ernest Cheesman, through his works, established that Linnaeus *Musa sapientum* and *Musa paradisiaca* were actually cultivars that descended from two wild seed-producing species, *Musa acuminata* and *Musa balbisiana* which were first reported by Luigi Aloysius Colla [8]. He recommended reclassification of banana plants into three distinct groups of cultivars – those exhibiting primarily the morphological characteristics of *Musa balbisiana*, those exhibiting primarily the morphological characteristics of *Musa acuminata*, and those with characteristics of morphological combination of the two [7].

Fig. (1). A tall variety of banana plant (*Musa balbisiana* Colla).

Fig. (2). A dwarf cultivar of banana plant.

Floral stalk or peduncle

Leave-blade

Leaf-petiole or midrib

Bunches of female
ovaries developed
to bananas

Trunk or pseudo-stem

Floral rachis

Flower bract

Male flowers

Banana heart
composed of bracts
with male flowers
inside

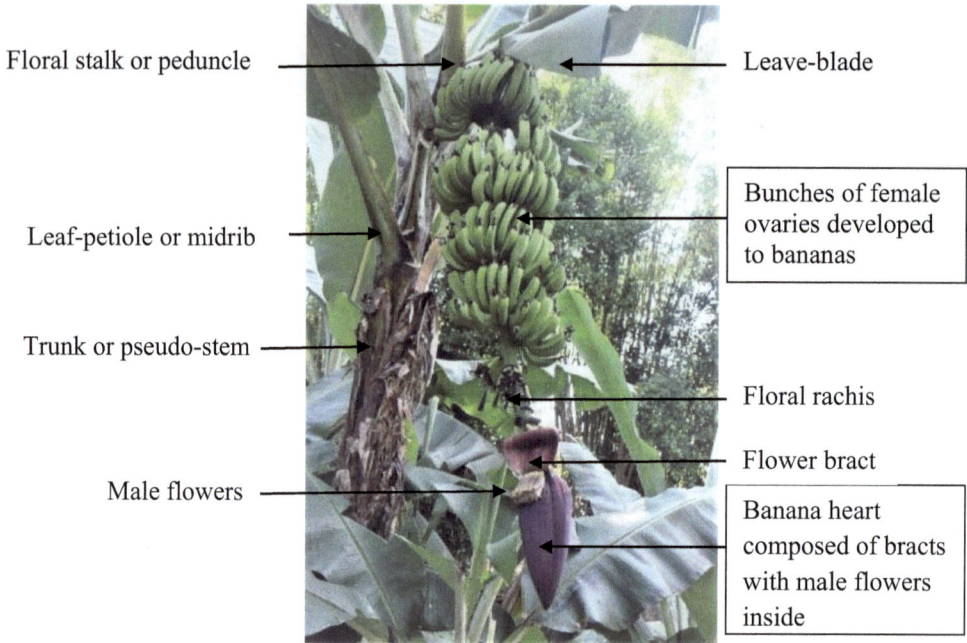

Fig. (3). A matured banana plant with a hanging cluster and inflorescence.

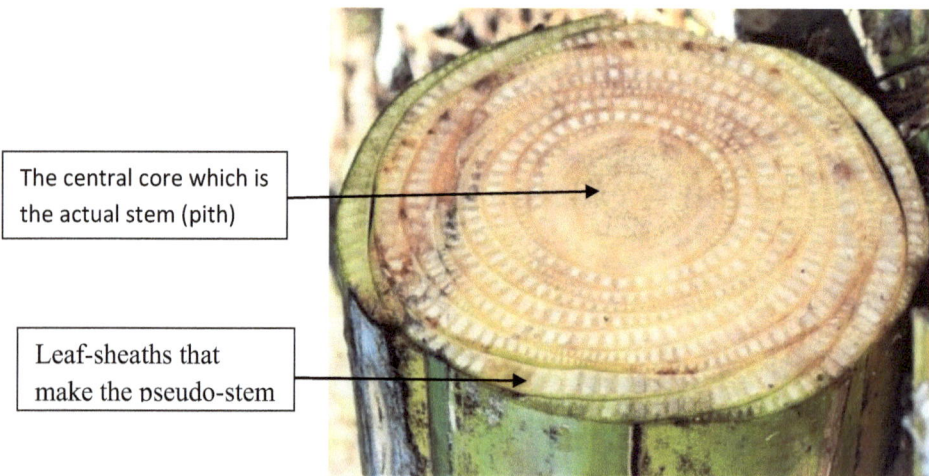

The central core which is
the actual stem (pith)

Leaf-sheaths that
make the pseudo-stem

Fig. (4). A cross-section of post-harvest banana trunk.

The family *Musaceae* accounts for about 50 different species of two genera –
Musa and *Ensete*. The genus *Musa* alone accounts for at least 35 species. Among
the 35 species *Musa paradisiaca* (plantain) and *Musa Sapientum* (dissert-banana)
are rich in starch [7]. The classification based on the number of chromosomes

divides the genus *Musa* into four sub-groups - *Australimusa, Callimusa, Eumusa,* and *Rhodochlamys*. Most of the ornamental species belong to the sub-groups *Callimusa* and *Rhodochlamys* [9]. Plantain and dessert banana cultivars belong to the sub-group *Eumusa*. These are natural hybrids of two wild species - *Musa acuminata* (contributing genome A) and *Musa balbisiana* (contributing genome B). Most of the domesticated bananas are triploid (2n=3x=33 chromosomes) among which dessert bananas mainly have genome constitution of AAA, and plantains have either AAB or ABB [10].

3. USES OF BANANA PLANT

Dessert bananas and plantains are important fruit crops for the populaces in tropical countries. These provide staple food for millions of people in developing countries. Although dessert bananas and plantains are perennial crops, they grow quickly and can be harvested throughout the year. In many tropical countries, plantains or cooking bananas are the main cultivars.

All parts of a banana plant find uses one way or the other - in food, feed, pharmaceutical, packaging, *etc*. Uses of banana leaves, fruits and sheaths for wound dressing in the ancient Egypt are reported [11]. Apart from leaves and fruits, whole banana plant is used in many social and religious ceremonies in India and some other countries.

3.1. Banana Flower

Banana flowers are popular and considered a healthy vegetable in many Asian countries. All the parts of the flower, including bracts are edible. These are cooked to prepare different cuisines such as soup, curry, fried dish, *etc.*

3.2. Banana Leaves

Banana leaves are suitable for use as eco-friendly wrapping material for preparing grilled or steamed foods as well as plates for serving foods. These are quite flexible as well as waterproof. In several Indian states such as Tamil Nadu, Karnataka, Andhra Pradesh and Kerala, serving food on a banana leaf is considered a healthy and auspicious tradition [12].

3.3. Banana Pseudo-stem

Traditionally the banana plant pseudo-stem is very popular for its rich fiber content [13]. Because of its rich edible fiber, the pseudo-stem as food is very beneficial for those who aim at weight loss. The central core of a matured banana trunk called pith is considered healthy and better as food than the pseudo-stem.

Japanese technology for the extraction of high quality textile fiber from banana trunk for clothing and household items dates back to the 13th century. Japanese used to cultivate banana farming exclusively for fiber – periodically they harvested soft leaves and shoots to ensure softness of fibers for yarn-making. They dyed and produced fibers of varying degrees of softness, yielding yarns with differing qualities for textiles of specific uses. The soft fibers were used to make traditional Japanese dresses like kimono and kamishimo [14]. In Nepal, traditional technology of mechanical fiber extraction, bleaching and dyeing was used to make hand-knitted rugs with silk-like texture. Biodegradable binding ropes can also be made from banana pseudo-stem fiber [15].

In South Indian states like Tamil Nadu, Kerala and Karnataka, post-harvest banana pseudo-stem is used to make fine threads for making flower garlands. Banana fiber and non-usable fruits are also used in the production of hand-made paper. Banana papers are usually hand-made in cottage industries and used in artistic works [16].

In North-Eastern states of India, '*kolakhar*' is traditionally being used for various purposes, especially by the rural folk. It is derived by extracting the ash obtained by open-air burning of the air-dried parts of banana plants or the peels of the ripe fruits. It is a traditional food additive [17] and known to help in normalizing digestive disorder of stomach. In Ayurvedic literature, *kolakhar* is known as *kadaliksāra* [18]. *Ksāra* means caustic alkali and *kadali* means banana plant. Thus, *kolakhar* or *kadaliksāra* means the caustic alkali derived from banana plant.

In ancient rural Assam and also in other North-Eastern states of India, *kolakhar* was widely used as soaps and detergents for washing cloths and shampooing hairs. After the markets are flooded with varieties of soaps and detergents, the use of *kolakhar* as a cleansing agent has been drastically reduced, yet in interior rural northeast its use for washing purposes still continues. It is reported that washing and cleansing with *kolakhar* help to grow and maintain long and healthy hair (Fig. 5) [19]. Many other uses of *kolakhar* are known in the rural northeast [20]. A few of them are:

  i. To prevent bacterial attack on freshly cut injuries. Application of *kolakhar* makes the healing faster.
  ii. To kill leeches and prevent their attack *kolakhar* is very useful for farmers while working in leech infested agricultural fields.
  iii. *Kolakhar* and *kolakhar* ash are used by farmers for cure and prevention of certain cattle diseases.

Fig. (5). A woman with healthy and long hair claimed to have achieved by using *kolakhar* instead of commercial soaps and shampoos [19].

In addition to the traditional uses, many modern day uses of *kolakhar* are possible. One such possible use is the isolation of potassium carbonate from *kolakhar*. Potassium carbonate has plenty of uses in industries including confectioneries, pharmaceutical industries, R & D laboratories, *etc*. But there is practically no natural source of potassium carbonate [21], and it is established that the major chemical component present in *kolakhar* is potassium carbonate [17]. Therefore, *kolakhar* can be a substitute for potassium carbonate in some uses and also can be a renewable source for the production of potassium carbonate. A laboratory process for the isolation of potassium carbonate from *kolakhar* has already been developed [22]. The process is yet to be tested for its commercial viability in large scale production. Another possible use of *kolakhar* is the isolation of potassium rich table salt [23]. Potassium rich table salt is known to help in balancing blood pressure [24 - 26].

A good number of modern day uses having potential for large scale commercial exploitation of *kolakhar* have been recently reported. *Kolakhar* ash has been tested in the laboratory as an excellent heterogeneous catalyst in the production of biodiesel from oils and fats [27 - 29]; it appears a potential catalyst for future biodiesel industries. Uses of *kolakhar* as catalysts or basic aqueous media to accomplish useful organic transformations have also been reported [30 - 33].

3.4. Banana

Although all parts of a banana plant find uses for various purposes, it is best known for its fruit, the bananas. According to FAOSTAT data of 2018, dessert banana with annual Global output of 115.74 million MT (13.38% of all fruits) tops the list of global fruits production [34, 35]. It along with plantains, serves as an ideal and low cost source of food in developing countries where most of the populaces depend on bananas for the supply of nutrients and minerals. Banana is a wholesome food, and the Ayurveda recommends bananas as the first solid food for babies [36].

4. MEDICINAL PROPERTIES OF BANANA AND BANANA PLANT

The banana plant and its fruits not only serve as the favorite food for people across the globe but also have incredible medicinal values. Not only fruits but also each and every part of a banana plant - the flowers, the stems, the leaf sheaths and the green leaves, all are equally useful for their medicinal values [37].

Young banana leaves can be used to treat burns, according to a report from Purdue University. The Department of Horticulture at Purdue University also asserts that banana leaves are a suitable remedy for diarrhea and effectively prevent the growth of ulcers [38]. The ashes of leaves and peels of unripe fruits are astringents and used for treating malignant ulcers. Green banana leaf juice without salt and sugar is believed to prevent various health problems such as bronchitis, nephritis, T.B, pleurisy, *etc.* It is also believed that the juice purifies blood and considered very effective against cough and cold, breathing problem, dropsy, constipation, dysentery, acidity, high BP, blood disorder, poisoning and liver problem. The chlorophyll in banana leaves is believed to provide protection against certain skin diseases, intestinal ulcer and leukemia.

The tender inner stem, also called the pith (Figs. **2** and **4**) of a matured banana plant, is used as food because of its medicinal values. Its high fiber content can help those who are on a weight-loss program. The use of banana stem as vegetable is believed to be beneficial to keep high blood pressure and diabetes under control. Banana stem and its juice are diuretic, detoxify our body, maintain fluid balance within the body and normalize stomach upset [39].

Extract of banana stem can help to dissolve the stones in the kidney and urinary bladder. It has been reported to suppress the formation of oxalate-associated kidney stones in animals, and may be a useful agent in the treatment of patients with hyperoxaluric calcium urolithiasis [40]. Houghton and coworkers have reported that the juice of banana pseudo-stem has antivenom action [41, 42].

In his book 'Herbal Foods and its Medicinal Values', H. Panda has described the mixture of banana plant pseudo-stem sap with coconut water as an effective treatment for nervous insomnia, epilepsy, dysentery, vomiting and hysteria [43]. In the book 'Speaking of Nature Cure', the authors have reported that the banana pseudo-stem juice clears the arteries, reduces blood cholesterol, and helps to fight urinary and digestive disorders [44].

Aziah *et al.* have advocated the use of native banana pseudo-stem flour (NBPF) and tender core of the banana pseudo-stem flour (TCBPF) in food applications to add calorie and edible fiber. NBPF and TCBPF are rich in fiber content. Their antioxidant activity, water holding capacity as well as oil holding capacity are considered beneficial for health [45]. Analgesic activity [46] and hepatoprotective effect [47] of the aqueous and ethanolic extracts of pseudo-stem of *Musa sapientum* Linn have also been reported. On the other hand, Benitez *et al.* have reported that the presence of mono potassium oxalate in the juice of banana pseudo-stem may cause muscular paralysis [48].

5. BANANA PRODUCTION IN THE WORLD

The banana plant is believed to be the oldest cultivated plant. It was probably domesticated first in Papua New Guinea, and is now considered native to tropical South and South-East Asian countries [49 - 51]. Bananas are now cultivated throughout the tropics, and grown in at least 135 countries across the globe. They are primarily cultivated for their fruits, and to a lesser extent to extract fiber, to prepare banana wine and to use certain ornamental varieties as decorative plants [52].

Based on the mode of uses, banana fruits are divided into two categories [7]:

1. Cooking bananas or plantains and
2. Dessert or sweet bananas.

The world banana market consists mainly of trade in dessert or sweet bananas where cavendish sub-group is prominent with a share of 47%. Dessert bananas contribute nearly 75% to the total banana production in the world. From small farms to large plantations, cavendish bananas are currently produced all over the world. India is the world's leading producer of cavendish bananas (26.33%) followed by China (9.65%), Indonesia (6.19%), Brazil (5.77%), Ecuador (5.43%) and Philippines (5.22%). These six countries together produce nearly 60% of global cavendish output. The top 10 banana producing nations are shown in Table **1**. Other types of banana produced in the world are highland banana, ABB and other cooking banana (24%), plantain AAB (17%), and Gros Michel and other dessert bananas (12%) [34, 53 - 55].

Table 1. Top 10 banana producing nations.

Banana Producing Countries (2017)			
Entry	**Country**	**Millions of Metric Tons**	**% of World Total**
1	India	30.48	26.33
2	China	11.17	9.65
3	Indonesia	7 .16	6.19
4	Brazil	6 .68	5.77
5	Ecuador	6.28	5.43
6	Philippines	6.04	5.22
7	Angola	4.30	3.72
8	Guatemala	3.89	3.36
9	Colombia	3.79	3.27
10	Tanzania	3.48	3.01
11	All other countries	32.47	28.05
	World total	115.74	100

6. BANANA PRODUCTION IN INDIA

Banana is one of the most important fruit crops in India. The banana culture in India is as old as Indian civilization. In India, Andhra Pradesh is the leading producer of banana, followed by Gujarat, Maharashtra, Tamil Nadu and Uttar Pradesh, each state having more than 10% share in the national production (Table **2**). Bananas are popular among all classes of people in India.

It is generally agreed that all the bananas and plantains are indigenous to the warm and moist regions of tropical Asia, comprising major parts of India, China, Burma and Thailand. The world's largest diversity in banana population is found in these regions. The banana production in different Indian states in the year 2017-18 is shown in Table **2** [53, 56].

Table 2. **Production of banana in India (top ten states) in the year 2017-18 [56].**

Entry	State	Production (x10^3 tons)	Share (%)
1	Andhra Pradesh	5003.07	16.24
2	Gujarat	4472.32	14.52
3	Maharashtra	4209.27	13.66
4	Tamil Nadu	3205.04	10.40
5	Uttar Pradesh	3172.33	10.30
6	Karnataka	2328.90	7.56
7	Madhya Pradesh	1834.03	5.95
8	Bihar	1396.39	4.53
9	West Bengal	1200.00	3.90
10	Kerala	1119.16	3.63
11	All other states	2866.99	9.31
	Total	30807.50	100

Source: National Horticulture Board (NHB), Govt. of India, 2018.

7. CHEMICAL COMPOSITION OF DIFFERENT MORPHOLOGICAL PARTS OF BANANA PLANT

Apart from fruits, other morphological parts of banana plants include pseudo-stem, foliage, tender core, floral stalk and rachis. Sweet banana and plantain together account for about 18% of the world's total fruit production [34]. These are ideal foods especially for weaning mother and infants. They are rich sources of carbohydrates, antioxidants, and minerals, especially potassium and iron. Banana peels are also rich in vitamins, sugars, pectins and lignins. These can be used as cattle feed and base material for biogas production, alcohol production and for protein extraction. Leaves are good source of lignocellulose. The central core of a matured plant (also called the pith, it is a stem in true sense) can be used as food and color absorber [57]. Because of its richness in different nutraceuticals, banana plant juice has potential uses in pharmaceutical industries.

The general chemical composition of different morphological parts of banana plant (Dwarf Cavendish) is reported [58]. Elemental analysis of ashes reveals the presence of potassium, calcium and silicon salts as major components. Analysis is important because inorganic elements have a negative effect on the Kraft pulping [59 - 61], chemicals and energy recovery [62, 63]. Inorganic elements have also bearing on paper quality and yield [64]. Their high content in banana plant deserves a special attention, especially when applied to pulping. This aspect may be considered a serious disadvantage in the use of banana plant as a raw material for pulp and paper production.

All morphological parts of banana plant contain considerable amount of ashes (12-27%) [58], which are considerably high when compared with corresponding data of other fast growing plants. The high ash content in floral stalk or rachis (about 27%) is noteworthy. Also equally noteworthy the high potassium contents of floral stalk, leaf sheaths and rachis. High potassium content of leaf sheaths or pseudo-stem bears significance as because it is the major morphological part of a banana plant.

8. IMPORTANCE OF POTASSIUM FOR PLANTS

Potassium electrolyte is an essential nutrient for the overall growth and proper reproduction of plants [65]. This essential nutrient is typically received by plants from soil in the form of potassium chloride. In fertilizer industry and commerce, it is marketed under the trade name of Muriate of Potash (MOP) which is essentially potassium chloride [66]. Potassium is essential for stimulating the growth of strong stems and to help the plant to develop disease resistance. It helps plants in regulating fluid balance, nerve signals and muscle contractions. It has vital role in the production of plant proteins and sugars. It also protects plants against draught by maintaining plants water content. Maintaining water content in plants is essential to help photosynthesis, to improve color, flavor and shelf-life of fruits and vegetables.

8.1. Natural Sources of Potassium

Potassium is abundantly found in nature including fresh fruits and vegetables. Almost all plants are rich in potassium, and among them banana plant is noteworthy [17, 67, 68]. Plants acquire potassium from soil as a nutrient for their growth. In agriculture, crops are provided potassium in the form of manures during cultivation. MOP is an important component of crop manures, and it provides potassium to plants as an electrolyte. It is a chemical fertilizer which mainly consists of potassium chloride. MOP or potassium chloride is mined from its ores.

8.2. Ores Rich in Potassium Chloride

Potassium chloride rich rock deposits (potash ores) are found in many regions of the earth's crust. These are the minerals derived from the ancient seas which dried up hundreds of thousands of years ago. These deposits are the sources of potash being mined today. The deposits are predominantly potassium chloride (KCl) along with sodium chloride (NaCl) as the minor component. As the time passed, these deposits were buried by thousands of feet of earth [69]. Potassium is the seventh most abundant element on the earth's crust but can't found in the elemental form. In the form of various compounds it constitutes 2.6% of the earth's crust. Corresponding abundance of sodium is approximately 1.8% [70]. In seawater, potassium is far less abundant (0.39 g/L) as compared to sodium (10.8 g/L) [71, 72].

Potassium minerals are mined and processed primarily to produce potassium chloride. In fertilizer industries, potassium chloride is often branded as potash, muriate of potash, or simply MOP. Potash for fertilizer is produced by processing potassium chloride rich rock deposits requiring only physical grading and separation of other minerals. Some of the potassium ores abundantly used for the production of MOP are sylvinite (KCl with NaCl), silvite (KCl), carnallite ($KCl.Mg.Cl_2.6H_2O$), kainite ($MgSO_4.KCl.3H_2O$), langbeinite ($MgSO_4.K_2SO_4$) and polyhalite ($K_2SO_4.MgSO_4.CaSO_4$) [73, 74].

8.3. Global Potash (Potassium Chloride) Production

Although huge volumes of potash ore reserves are currently available, these are confined only to a few countries of the world. Major potash producing countries are Canada, Belarus, Russia and China (Table **3**). Global potash production in 2018 was estimated at 68.1 million tones of which 33% is produced by Canada alone. Naturally Canada is the world's largest potash producer as of now. Canada, Belarus, Russia and China together account for nearly 80% of the world's total potash production [75, 76]. Potash is also produced by Germany, Israel, United States, Jordan, and a few more countries [77, 78]. In Carlsbad, New Mexico, potash is extensively produced from sylvinite, a sodium chloride - potassium chloride mineral. Apart from rock deposits, potash is also recovered along with other compounds from the brine of Searles Lake in California and from the lake brines in Utah, USA. World's largest potash producing countries and their estimated reserves are shown in Table **3**.

Table 3. Country wise production of potash (potassium chloride) 2018 and total reserves.

Producer Ranking	Country	Tonnes (in Thousands)	Percentage of Total Production	Reserves in Potassium Oxide Equivalent (Million Tonnes)	Percentage of Total Reserve
1	Canada	22,680	33.3	1,200	20.6
2	Belarus	12,043	17.7	750	12.9
3	Russia	11,753	17.3	2,000	34.3
4	China	7,382	10.8	350	6.0
-	Other countries	14,256	20.9	1525	26.2
Total		68,114	100.0	5,825	100.0

8.4. Consumers of Potash (Potassium Chloride)

Nearly 95% of potash is used as a fertilizer to increase crop yield and support plant growth. Only small quantities are used in the manufacture of potassium-bearing chemicals for use in research laboratories and in industries to produce detergents, ceramics, pharmaceuticals, as alternative to de-icing salt, *etc.*

As compared to the supply environment, global demand environment for potash is much more balanced [79]. South and East Asia, Latin and North America, and west and central Europe are the major consumers of potash [80]. Two-thirds of world total potash is consumed by four countries - China, the United States, Brazil, and India [81]. Brazil imports 90% of the potash it consumes, and India imports 100%.

Although known potash reserves of the world are considered adequate to meet the future demand environment for many years to come, the current deposits are non-renewable, and owned by a few countries only. Although Asian regions accounted for about 50% of the world K_2O consumption in 2018-2019, but by region it produced potash less than 15% of its total needs. The current world potash consumption is small as compared to the estimated reserves, but consumption is growing faster than expected. The United States produces and consumes a heavy amount of potash, but the prospect for an adequate supply by the country are not so bright as compared to the growing demand in that country. Currently potash is being extensively used only by a few country, many countries are yet to use potash as fertilizer even now. But these countries are expected to use potash in near future to afford more food to meet the demand by growing population. If these countries continue to increase potash consumption by extensive use in farming, then the world's potential demand for potash is expected to increase by many-fold. Considerable care is therefore needed to protect the existing reserves against unwise use. Additional research is needed to explore newer sources of potash or to advance technical skills to recover potash from known low-grade sources. In this context, banana plant may be a cheap, renewable and easily available alternative source of potassium chloride. Almost all morphological parts of banana plant are rich in potassium chloride. Therefore, isolation of potassium chloride from banana plant or finding uses of banana plant to replace potash would be an innovative idea.

9. POTASSIUM CARBONATE

Potassium carbonate finds wide ranges of applications including applications in industries. Some of the industrial applications include applications (i) in the production of soaps and glasses, (ii) as a mild drying agent, (iii) a buffering agent in the production of mead and wine, (iv) in hard water softening, (v) as a deep-fat fryer fire extinguisher, (vi) in condensed aerosol fire suppression, (vii) an ingredient in the fluxes for welding and arc-welding rods, (viii) as a stable source of energy in gunpowder, *etc.*

There is virtually no source of potassium carbonate in nature except a few minor sources like a few African lakes and Dead Sea [21]. In the olden days prior to industrial era, potassium carbonate was prepared by aqueous extraction of wood ashes and then evaporating the resulting solution in large iron pots to yield a white residue called 'pot ash'. Approximately 10% by weight of common wood ash could be recovered as pot ash [81, 82]. Later, 'pot ash' became 'potash' and the term is now widely applied to naturally occurring potassium salts and the commercial products derived from them [83, 84].

In ancient Europe, production of potash led to large scale destruction of natural vegetation [85]. Nearly 500 kg of vegetation was required for the production of less than 1 kg of potash. Since the middle of the nineteenth century, the saline residues and the salt deposits of rock salt industries have been used as the raw materials for the production of potassium carbonate [86].

Currently most preferred industrial process for the production of potassium carbonate involves the carbonation of electrolytically produced potassium hydroxide [85, 87]. In this process, potassium hydroxide is first prepared by electrolysis of KCl. Saturation of potassium hydroxide solution with carbon dioxide followed by partial evaporation leads to precipitation of hydrated potassium carbonate $K_2CO_3.1.5H_2O$. Anhydrous potassium carbonate from hydrated carbonate can be prepared by calcination at 250 to 350 °C.

$$2KOH + CO_2 \rightarrow K_2CO_3 + H_2O$$

In another process, potassium hydrogen carbonate is first prepared by reacting potassium chloride with carbon dioxide under pressure in a precarbonated isopropylamine solution. Calcination of potassium hydrogen carbonate affords potassium carbonate [88]. A serious environmental problem associated with this process is the disposal of the contaminated calcium chloride brine which is produced as a byproduct. In the former states of the USSR, potassium carbonate is produced as a byproduct along with the production of aluminium hydroxide by the nepheline decomposition process [21].

Due to the extensive use of potassium carbonate in different fields, including industrial applications, the demand for potassium carbonate is increasing day by day. But the production of potassium carbonate is limited as practically no source of K_2CO_3 is available in nature. Production from wood ash derived from the destruction of vegetation is not only unaffordable but also economically unviable. Production of K_2CO_3 by absorption of CO_2 in KOH solution is an energy intensive process. So the cost of production of K_2CO_3 becomes high and therefore, a renewable and ready source of potassium carbonate is highly desirable. It has been reported that *kolakhar,* which is derived from different parts of banana plant, is a rich source of K_2CO_3 and therefore, can be commercially exploited as a cheap and renewable natural source of K_2CO_3 [17, 22]. After harvesting, banana plants have practically no use for farmers and are simply thrown away. So, it can be a novel idea to explore the use of post-harvest banana plants as an alternative fast growing renewable natural source for the production of K_2CO_3.

To sum up, banana tops the list of global fruits in terms of production contributing about 13.38% of the world's total fruits production, and they are cultivated throughout the tropics. Food and Agriculture Organization of the United Nations reported that about 155.22 million metric tons of bananas (including plantains) were produced all over the world in the year 2018, and India is a leading nation in banana production. According to the National Horticulture Board of India, banana cultivated area in India is increasing day by day. It was 830000 hectares in India in the year 2011, in 2018 it has recorded 884000 hectares. Each hectare of banana crop generates nearly 250 tons of plant residual wastes and each ton of banana fruits generates waste material 7 to 8 times of its weight. Transportation of these materials from farm lands to elsewhere is economically unviable for farmers. Therefore, disposal of the residual wastes by farmers into nearby rivers, lakes and on roads has become a serious environmental concern.

The main residual wastes of banana crop are the pseudo-stems. Although banana pseudo-stem has some minor uses such as uses in social and religious functions, fiber production, preparation of '*khar*' (or *kolakhar,* a traditional food additive in Assam) in Assam and other north east states, no major use of banana plant pseudo-stem has been reported. The production of banana fruits is huge and it generates pseudo-stems as wastes in millions of tons. So, finding a novel application of banana plant pseudo-stem is important to bring additional benefits to banana farmers.

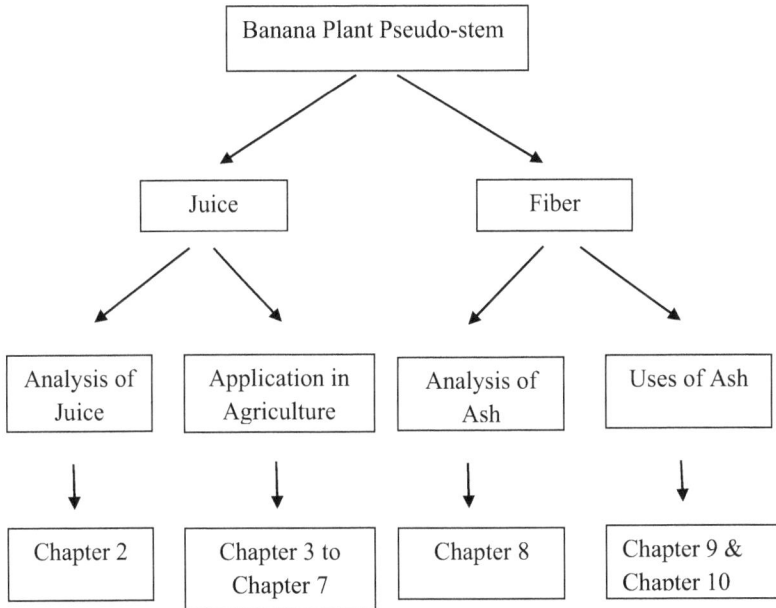

```
              Banana Plant Pseudo-stem
                 /              \
              Juice            Fiber
              /    \           /     \
  Analysis of  Application  Analysis  Uses of Ash
    Juice      in Agriculture  of Ash
     |            |            |          |
  Chapter 2   Chapter 3 to  Chapter 8  Chapter 9 &
              Chapter 7                 Chapter 10
```

The objective of this book is to show how banana plant pseudo-stem juice can be used in crops cultivation to supplement potash and replace chemical manure, the muriate of potash as well as to show how the fiber component of the pseudo-stem can be used to produce potassium rich table salt and potassium rich carbonate. At a glance, the book is about the agricultural use of postharvest banana plant pseudo-stem to inculcate value addition to banana farming.

REFERENCES

[1] Stover, R.H.; Simmonds, N.W. *Bananas,* 3rd ed; Longman Scientific & Technical: Harlow, Essex, England; Wiley: New York, **1987**.

[2] *Office of the Gene Technology Regulator. The Biology of Musa L. (Banana), Version 1*; Department of Health and Ageing, Australian Government, **2008**.

[3]   Frame, P. *General Information about K-40.*, **1999**, www.orau.org/ptp/collection/consumer products/potassiumgeneralinfo.htm

[4]   Edwards, G. *About Radioactive Bananas,* www.ccnr.org/About_Radioactive_Bananastinyurl.com /kgcm3ywtinyurl.com/kgcm3yw

[5]   GRIN. *Genus: Musa L. In: Taxonomy for Plants*; National Genetic Resources Program **2009**.

[6]   Carl, L. *Species Plantarum (in Latin for The Species of Plants)*; Laurentius Salvius: Stockholm, Sweden, **1753**.

[7]   Valmayor, R.V.; Jamaluddin, S.H.; Silayoi, B.; Kusumo, S.; Danh, L.D.; Pascua, O.C.; Espino, R.R.C. *Banana Cultivar Names and Synonyms in Southeast Asia*; International Network for the Improvement of Banana and Plantain (INIBAP): Montpellier: France, **2000**.

[8]   *Musa paradisiaca,* www.users.globalnet.co.uk

[9]   Aparna, V. Ornamental Bananas. *Vikaspedia,* **2019**. Accessed on 2020-04-27.

[10]   Heslop-Harrison, J.S.; Schwarzacher, T. Domestication, genomics and the future for banana. *Ann. Bot.,* **2007**, *100*(5), 1073-1084. [http://dx.doi.org/10.1093/aob/mcm191] [PMID: 17766312]

[11]   Roberts, M.J. *Edible and Medicinal Flowers,* 1 sted; The Spearhead Press (An imprint of New Africa Books (Pty) Ltd.): Claremont, **2000**.

[12]   Banan Leaves: Traditional way of eating. In: *Wikipedia the free enclypedia*; Tamilnadu Tourism, **2020**.

[13]   Simple Indian Recipes *SimpleIndianRecipes.com,* **2018**. Retrieved on 2020-05-09.

[14]   Kijoka banana fiber cloth (Kijoka no bashofu). *Wikipedia the free encyclopedia,* **2015**, Accessed on 2020-04-28.

[15]   *The Encyclopedia of Crafts in Asia Pacific Region (APR): Traditional Handmade Products,* https://encyclocraftsapr.com/

[16]   Singh, L.; Bandyopadhyay, T.K. Handmade paper from banana stem. *IJSIR,* **2013**, *4*(7), 1074-1079.

[17]   Deka, D.C.; Talukdar, N.N. Chemical and spectroscopic investigation of *kolakhar* and its commercial importance. *Indian J. Tradit. Knowl.,* **2007**, *6*(1), 72-78.

[18]   Reddy, K.R.C. *Bhaisajya KalpanāVijñānam (A Science of Indian Pharmacy),* 2nd ed; Chaukhambha Orientalia: Sanskrit Bhawan: Varanasi, **2001**, p. 350.

[19]   Staff report on use of Kolakhar. *Āmār Asom (an Assamese daily),* **2002**, July 13.

[20]   *Direct information from people on 'Traditional knowledge practiced in rural Assam',* **2015**.

[21]   *Ullman's Encyclopedia of Industrial Chemistry,* 6th ed; Wiley-VCH: Germany, **2003**, Vol. 29, pp. 93-160.

[22]   Deka, D.C.; Talukdar, N.N. A method for the isolation of potassium carbonate from banana plants. Indian Patent No.222135, 2008.

[23]   Neog, S.R.; Deka, D.C. Salt substitute from banana plant (*Musa-balbisiana* Colla). *J. Chem. Pharm. Res.,* **2013**, *5*(6), 155-159.

[24]   *Dietary reference intakes for water, potassium, sodium, chloride and sulfate,* **2004**, www.nap.edu/ Accessed on 2020-02-08.

[25]   Barri, Y.M.; Wingo, C.S. The effects of potassium depletion and supplementation on blood pressure: a clinical review. *Am. J. Med. Sci.,* **1997**, *314*(1), 37-40. [PMID: 9216439]

[26]   Hajjar, I.M.; Grim, C.E.; George, V.; Kotchen, T.A. Impact of diet on blood pressure and age-related

changes in blood pressure in the US population: analysis of NHANES III. *Arch. Intern. Med.,* **2001**, *161*(4), 589-593.
[http://dx.doi.org/10.1001/archinte.161.4.589] [PMID: 11252120]

[27] Deka, D.C.; Basumatary, S. High quality biodiesel from yellow oleander (*Thevetia peruviana*) seed oil. *Biomass Bioenergy,* **2011**, *35*, 1797-1803.
[http://dx.doi.org/10.1016/j.biombioe.2011.01.007]

[28] Basumary, S.; Deka, D.C.; Deka, D.C. Composition of biodiesel from *Gmelina arborea* seed oil. *Adv. Appl. Sci. Res.,* **2012**, *3*(5), 2745-2753.

[29] Basumatary, S.; Barua, P.; Deka, D.C. *Gmelina arborea* and *Tabernaemontana divaricata* seed oils as non-edible feedstocks for biodiesel production. *Int. J. Chemtech Res.,* **2014**, *6*(2), 1440-1445.

[30] Dwivedi, K.D.; Borah, B.; Chowhan, L.R. Ligand free one-pot synthesis of pyrano[2,3-*c*]pyrazoles in water extract of banana peel (WEB): A green chemistry approach. *Front Chem.,* **2020**, *7*, 944.
[http://dx.doi.org/10.3389/fchem.2019.00944] [PMID: 32039156]

[31] Sarmah, M.; Mondal, M.; Bora, U. Agro-waste extract based solvents: Emergence of novel green solvent for the design of sustainable processes in catalysis and organic chemistry. *ChemistrySelect,* **2017**, *2*, 5180-5188.
[http://dx.doi.org/10.1002/slct.201700580]

[32] Baruah, P.R.; Ali, A.A.; Saikia, B.; Sarma, D. A protocol for ligand free Suzuki–Miyaura cross-coupling reactions in WEB at room temperature. *Green Chem.,* **2015**, *17*(3), 1442-1445.
[http://dx.doi.org/10.1039/C4GC02522A]

[33] Konwar, M.; Ali, A.A.; Sarma, D. A green protocol for peptide bond formation in WEB. *Tetrahedron Lett.,* **2016**, *57*(21), 2283-2285.
[http://dx.doi.org/10.1016/j.tetlet.2016.04.041]

[34] *FAO* ©*Statista ,* **2020**. Accessed on 2020-02-11.

[35] *World Bananas Production,* **2018**. knoema.com, Accessed on 2020/04/29.

[36] Suri, S. *Ayurveda Medicinal Properties of Banana.,* **2012**. https://EzineArticles.com/expert /Dr._Savitha _Suri/64291

[37] Kumar, K.P.S.; Bhowmik, D.; Duraivel, S.; Umadevi, M. Traditional and medicinal uses of banana. *J. Pharmacogn. Phytochem.,* **2012**, *1*(3), 51-63.

[38] King, T. Uses of Banana Leaves. *Garden Guides,* https://www.gardenguides.com/13428204-uses--f-banana-leaves.html

[39] Prasad, K.V.S.R.G.; Bharathi, K.; Srinivasan, K.K. Evaluation of Musa (Paradisiaca Linn. Cultivar) – 'Puttubale' stem juice for antilithiatic activity in albino rats. *Indian J. Physiol. Phannaco.,* **1993**, *37*(4), 337-341.

[40] Rajesh, N. Medicinal benefits of *Musa paradisiaca* (Banana). *Int. J. Biol. Res.,* **2017**, *2*(2), 51-54.

[41] Houghton, P.J.; Osibogun, I.M. Flowering plants used against snakebite. *J. Ethnopharmacol.,* **1993**, *39*(1), 1-29.
[http://dx.doi.org/10.1016/0378-8741(93)90047-9] [PMID: 8331959]

[42] Houghton, P.J.; Skari, K. The effect of Indian plants used against snakebite on blood clotting. *J. Pharm. Pharmacol.,* **1992**, *44*, 1054-1060.

[43] Panda, H. *Herbal Foods and its Medicinal Values*; National Institute of Industrial Research: Delhi, **2003**.

[44] Lakshmana, K.; Swaminathan, S. *Speaking of Nature Cure*; Sterling Publishers Pvt. Ltd: Greater Noida, **2011**.

[45] Aziz, N.A.A.; Ho, L-H.; Azahari, B.; Bhat, R.; Cheng, L-H.; Ibrahim, M.N.M. Chemical and functional properties of the native banana (*Musa acuminata×balbisiana* Colla *cv.* Awak) pseudo-stem

and pseudo-stem tender core flours. *Food Chem.,* **2011**, *128*(3), 748-753.
[http://dx.doi.org/10.1016/j.foodchem.2011.03.100]

[46]   Ingale, S.P.; Ingale, P.L.; Joshi, A.M. To study analgesic activity of stem of *Musa sapientum* Linn. *J. Pharm. Res.,* **2009**, *2*(9), 1381-1382.

[47]   Dikshit, P.; Tyagi, M.K.; Shukla, K.; Sharma, S.; Gambhir, J.K.; Shukla, R. Hepatoprotective effect of stem of *Musa sapientum* Linn in rats intoxicated with carbon tetrachloride. *Ann. Hepatol.,* **2011**, *10*(3), 333-339.
[http://dx.doi.org/10.1016/S1665-2681(19)31546-7] [PMID: 21677336]

[48]   Benitez, M.A.; Navarro, E.; Feria, M.; Trujillo, J.; Boada, J. Pharmacological study of the muscle paralyzing activity of the juice of the banana trunk. *Toxicon,* **1991**, *29*(4-5), 511-515.
[http://dx.doi.org/10.1016/0041-0101(91)90025-M] [PMID: 1862523]

[49]   Nayar, N.M. The bananas: botany, origin, dispersal. In: *Horticultural Reviews*; Janick, J., Ed.; Wiley-Blackwell: USA, **2010**. Vol. 36. 10.1002/9780470527238.ch2.

[50]   De Paz-Sànchez, M. 'Plantain of Guinea'. The Atlantic adventure of banana. *GJSFR: C Biol. Sci.,* **2014**, *14*(2), 49-65. version 1.0

[51]   Sardos, J.; Christelová, P.; Čížková, J.; Paofa, J.; Sachter-Smith, G.L.; Janssens, S.B.; Rauka, G.; Ruas, M.; Daniells, J.W.; Doležel, J.; Roux, N. Collection of new diversity of wild and cultivated bananas (*Musa* spp.) in the Autonomous Region of Bougainville, Papua New Guinea. *Genet. Resour. Crop Evol.,* **2018**, *65*, 2267-2286.
[http://dx.doi.org/10.1007/s10722-018-0690-x]

[52]   *Pro-Musa: The Banana Knowledge Platform,* www.promusa.org/Banana-producing

[53]   *Horticultural Statistics at a Glance*; Government of India, Department of Agriculture, Cooperation & Farmers' Welfare, Horticulture Statistics Division: New Delhi, **2018**.

[54]   *Banana Market Review – Preliminary Results*; FAO of United Nations.: Rome, Italy, **2019**.

[55]   Maghuyop, M.A.; Borromeo, K., Eds. Advancing banana and plantain R & D in Asia and the Pacific; A.B. Molina, V.N. Roa, I. Van den Bergh Maghuyop, M.A.; Borromeo, K., Eds. *Proceedings of the 2nd BAPNET Steering Committee,* 6-11 October 2003. Montpellier, France **2004**, *12*.

[56]   *Agri Xchange-APEDA: Indian Production of Banana, 2017-18.* www.apeda.gov.in

[57]   Alwi, H.; Idris, J.; Musa, M.; Hamid, K.H.K. A preliminary study of banana stem juice as a plant-based coagulant for treatment of spent coolant wastewater. *J. Chem.,* **2013**, 1-7.
[http://dx.doi.org/10.1155/2013/165057]

[58]   Oliveira, L.; Cordeiro, N.; Evtuguin, D.V.; Torres, I.C.; Silvestre, A.J.D. Chemical composition of different morphological parts from 'Dwarf Cavendish' banana plant and their potential as a non-wood renewable source of natural products. *Ind. Crops Prod.,* **2007**, *26*, 163-172.
[http://dx.doi.org/10.1016/j.indcrop.2007.03.002]

[59]   Higuchi, T. *Biochemistry and Molecular Biology of Wood,* 1st ed; Timell, T.E., Ed.; Springer Series in Wood Science: Springer-Verlag: Berlin Heidelberg, **1997**.
[http://dx.doi.org/10.1007/978-3-642-60469-0]

[60]   Rahmana, M.M.; Nayeema, T.I.J.; Jahana, M.S. Variation of chemical and morphological properties of different parts of banana plant (*Musa paradisica*) and their effects on pulping. *Int. J. Lignocellulos. Prod.,* **2014**, *1*(2), 93-103.

[61]   Obernberger, I.; Biedermann, F.; Widmann, W.; Riedl, R. Concentrations of inorganic elements in biomass fuels and recovery in the different ash fractions. *Biomass Bioenergy,* **1997**, *12*(3), 211-224.
[http://dx.doi.org/10.1016/S0961-9534(96)00051-7]

[62]   Keitaanniemi, O.; Virkola, N.E. Undesirable elements in causticizing systems. *Tappi J.,* **1982**, *65*(7), 89-92.

[63]   Taylor, K. Effect of Non-Process Elements on Kraft Mill Efficiency. *PAPTAC Parksville Conference,* April 20-21**2007**.https://www.researchgate.net/publication/228873894

[64]   Jeyasingam, J.T. A summary of special problems and considerations related to non-wood pulping worldwide. *In: TAPPI Pulping Conference,* New Orleans, LA, USA, **1988**, *III*, pp. 571-579.

[65]   Prajapati, K.; Modi, H.A. The importance of potassium in plant growth-A review. *Ind. J. Plant Sci.,* **2012**, *1*(02-03), 177-186.

[66]   Hasanuzzaman, M.; Borhannuddin Bhuyan, M.H.M.; Nahar, K.; Hossain, S.; Mahmud, J.A.; Hossen, S.; Masud, A.A.C. Potassium: A vital regulator of plant responses and tolerance to abiotic stresses. *Agronomy,* **2018**, *31*, 29.

[67]   Deka, D.C.; Talukdar, N.N. *Kolakhar and its Chemistry,* 2nd ed; Techno Ed. Publication: Guwahati, **2011**.

[68]   Deka, D.C.; Talukdar, N.N. Kolakhar and its Chemistry.*Appropriate Technologies for North East India*; Associated Publishing Company: New Delhi, **2012**.

[69]   Fite, R.C. Origin and occurrence of commercial potash deposits. *Proc. Okla. Acad. Sci.,* **1951**, *32*, 123-125.

[70]   Greenwood, N.N.; Earnshaw, A. *Chemistry of the Elements,* 2nd ed; Butterworth-Heinemann© Elsevier Ltd, **1997**.

[71]   Webb, D.A. The sodium and potassium content of sea water. *J. Exp. Biol.,* **1939**, *16*, 178-183.

[72]   Micale, G.; Cipollina, A.; Rizzuti, L. Seawater desalination for freshwater production. In: *Seawater Desalination: Conventional and Renewable Energy Processes*; Cipollina, A.; Micale, G.D.M; Rizzuti, L., Eds.; Springer.Com, **2009**; pp. 1-15. Ch.1. [http://dx.doi.org/10.1007/978-3-642-01150-4_1]

[73]   Norman, L.; Weiss, , Eds. *SME Mineral Processing Handbook*; In-Chief; Society of Mining Engineers of the American Institute of Mining, Metallurgical, and Petroleum Engineers, Inc.: New York, **1985**.

[74]   Garrett, D.E. *Potash: Deposits, Processing, Properties and Uses*; Springer: Netherlands, **1996**. [http://dx.doi.org/10.1007/978-94-009-1545-9]

[75]   *World Fertilizer Trends and Outlook to 2020: Summary Report; FAO*; United Nations: Rome, Italy, **2017**.

[76]   *Potash Facts- Natural Resources Canada,* **2019**. https://www.nrcan.gc.ca/potash-facts

[77]   *Review of Potassium Chloride Market in the CIS,* **2004**, www.infomine.ru

[78]   Mineral Commodity Summaries. *US Geological Survey,* **2019**.

[79]   Executive, S.F.O. 2019 – 2023; International Fertilizer Association (IFA) *IFA Annual Conference,* Montreal (Canada) **2019**.

[80]   *World Fertilizer Trends and Outlook to 2018,* **2015**, www.fao.org/publications

[81]   *Potash in Wikipedia,* **2015** , Accessed on 2020-05-03.

[82]   Harper, D. *Potash In: Online Etymology Dictionary,* https://www.etymonline.com/word/potash

[83]   Prud'homme, M. *Potash In: The Canadian Encyclopedia,* **2015**, https://www.thecanadian encyclopedia.ca/en/article/potash

[84]   Brasnett, R. Outlook for Potash *Proceedings of the 50th Anniversary Annual Meeting, Fertilizer Industry Roundtable,* October 4 New Orleans: Louisiana **2000**.

[85]   Potassium compounds. In: *Kirk- Othmer Encyclopedia of Chemical Technology,* 5th ed; Kirk, Othmer, Ed. John Wiley & Sons: New York, **2007**; p. 20.

[86]   Buchel, K.H.; Moretto, H-H.; Woditsch, P. *Industrial Inorganic Chemistry,* 2nd ed; Terrell, D.R., Ed.;

Wiley-VCH: New York, **2000**.
[http://dx.doi.org/10.1002/9783527613328]

[87]   Schultz, H.; Bauer, G.; Schachl, E.; Hagedorn, F.; Schmittinger, P. Potassium compounds. In: *Ullmann's Encyclopedia of Industrial Chemistry*; Wiley-VCH: Weinheim, **2012**; 29, pp. 639-704.

[88]   Oates, J.A.H. Potassium carbonate. In: *Lime and Limestone: Chemistry and Technology, Production and Uses*; Wiley-VCH: Weinhein, New York, **1998**.

CHAPTER 2

Analysis of Banana Plant (*Musa balbisiana* Colla) Pseudo-Stem Juice

Abstract: Materials and methods for the extraction and analysis of banana plant pseudo-stem juice have been discussed. The pseudo-stem consists of nearly 95% juice and less than 5% fibers. The juice is quite rich in potassium and oxalate, moderately rich in sodium and chloride. In addition, nitrate, phosphate and a couple of heavy metals in trace concentrations have been detected. Two important bioactive organic molecules, namely (*E*)-4- (4-methoxyphenyl) but-3-en-2-one and (1*E*,4*E*)-1, 5-bis-4-methoxyphenyl)penta-1,4-dien-3-one, have been isolated from the juice, and characterized.

Keywords: (1*E*,4*E*)-1, 5-Bis (4-methoxyphenyl) penta-1, 4-dien-3-one, (*E*) -4-(4-Methoxyphenyl)but-3-en-2-one, Banana plant juice, Banana plant pseudo-stem.

1. INTRODUCTION

Banana plant juice is a transparent fluid which runs out from the pseudo-stem when cut and it represents 90-95 percent of the pseudo-stem weight [1, 2]. The juice becomes pink when exposed to air and after some time changes to light brown. The pseudo-stem juice is not only important for its medicinal values but also for its other applications.

Banana stem juice is beneficial to health, according to Ayurveda [3, 4]. To make the juice, chopped banana stem and water are ground until it becomes homogeneous. Banana stem juice combined with buttermilk and taken on an empty stomach helps in weight reduction [5]. The juice also heals ulcers, reduces burning sensation and acidity. Its astringent quality helps in blood coagulation. Banana stem is believed to have a cooling effect on the body and hence, is recommended in tropical climates.

Fresh pseudo-stem juice of *Musa sapientum* has a significant anticonvulsant activity, which might be due to the antioxidant potential of phyto-constituents present in it [6]. Herbal formulatation named as MTEC consisting of aqueous methanolic extracts of *Musa paradisiaca, Tamarinduss indica, Eugenia jambolana and Coccinia indica* is reported to have significant control over fasting blood glucose and serum insulin levels along with the testicular function in *Streptozocin* induced diabetic rat [7].

Banana pseudo-stem (BPS) was found to be a potential source of polyphenols, which can be used as natural antioxidants in food, nutraceutical, and pharmaceutical industries [8]. The multiple antioxidant properties may be an impetus to increase the consumption of BPS either in fresh or in processed form. The total phenolics and total flavonoids in various solvent extracts of pseudo-stem of different banana cultivars vary from 7.58 to 291 mg gallic acid equivalent (GAE/g of extract) and from 4 to 80 mg catechin equivalent (CE/g of extract), respectively. Acetone extract shows high antioxidant activity, whereas methanol extract exhibits high metal chelating activity [9]. Antioxidants, when present in food or in the body at low concentrations, markedly retard or prevent the oxidation of that substrate [10]. Uses of synthetic antioxidants are restricted due to their carcinogenicity [11]. Hence, interest has increased in finding naturally occurring antioxidants to replace synthetic antioxidants [12, 13]. Natural antioxidants have the capacity to improve food quality and can also act as nutraceuticals to terminate free radical chain reactions in biological systems. These benefits have been attributed to some of the phytochemical constituents and, in particular to polyphenols [14]. Preliminary phytochemical analysis reveals the presence of vitamin B, oxalic acid, vitamin C, tannin, glycosides, phenolic compounds, gum *etc.* in fresh stem juice of *Musa sapientum*. It is reported that *Musa sapientum* juice prevents the convulsions possibly through the prevention of inhibition of vitamin B_6 metabolism. The anti-convulsion effect of fresh stem juice of *Musa sapientum* might be due to antioxidant potential of phyto-constituents present in it [6].

The juice of the banana plant pseudo-stem produces a non-depolarizing neuromuscular block [15]. It is possible that the juice contains Acetylcholine (Ach, a neurotransmitter) antagonist that binds to Ach receptors in an irreversible way.

Banana plant juice is reported to help steel reinforcement to sustain the corrosive effects of aggressive ions in the environment surrounding the concrete when the latter has been mixed with banana plant juice. It is believed that the banana plant juice forms a protective layer on the concrete steel surface and hence prevents it from corrosion [1]. The effectiveness of banana stem juice as a natural coagulant for the treatment of spent coolant wastewater has also been investigated. The percentage of chemical oxygen demand (COD), suspended solids (SS), and turbidity removal by using banana stem juice have shown tremendous potential as a plant-based natural coagulant in the treatment of spent coolant waste water [16].

2. MATERIALS

Amongst the different species of banana plants available in Assam, the trunk of *Musa balbisiana* (*Bhimkol* or *Athiakol* in Assamese), the seedy variety of banana plant has been selected for this experiment because the plant of this species grows easily in Assam, has resistance to diseases and flood, nutritious fruit and a stout trunk. The post-harvest banana plant (*Musa balbisiana*) was randomly selected and collected from a household at Dhakuakhana, district Lakhimpur, Assam. The plant specimen was authenticated by the Department of Botany, Dhakuakhana College.

3. METHODOLOGY

3.1. Separation of Fresh Pseudo-stem Juice of Banana Plant

A fresh piece of pseudo-stem of post-harvest banana plant was taken and the leaf sheaths and tender core (floral stalk) were manually separated from the pseudo-stem. The separated leaf sheaths and tender core were washed well in running water. Juices were extracted from the leaf sheath and the tender core separately with the help of a sugar-cane juicer machine by squeezing again and again to extract the juice as much as possible. The volumes of the juices, after filtration to remove suspended solids, were measured and kept in air tight containers for chemical and spectroscopic investigation. The juices were preserved in a refrigerator at 7 °C to ensure freshness. The fibers were dried under sun and weighed.

3.2. Measurement of pH of Juice Samples

The pH of the juice samples was recorded with the help of a digital pH meter (Eutech Instrument, pH 510, pH/mV/°C meter). Procedure given in the manual was followed. The meter was calibrated using buffer capsules of pH 4, pH 7 and pH 10. The cell of the meter was directly immersed in the experimental solution and pH was recorded at ambient temperature (24 °C).

4. METHODS OF INORGANIC ANALYSIS

4.1. Procedures Involved in Chemical and Spectroscopic Investigation on Juice

Qualitative analysis of the juice samples was done by the standard procedure of chemical tests [17] and the presence of the following acid and basic radicals as the major components were confirmed.

Acid radicals: Oxalate ($C_2O_4^{2-}$), Chloride (Cl^-), Nitrate (NO_3^-) and Phosphate (PO_4^{3-})

Basic radicals: Sodium (Na^+) and Potassium (K^+)

The concentrations of potassium and sodium were determined by flame photometry, chloride and oxalate by gravimetric analysis, and nitrate and phosphate by UV-Visible spectrophotometry. The concentrations of trace metals were determined by Atomic Absorption Spectroscopy.

4.2. Estimation of Oxalate Ion ($C_2O_4^{2-}$)

The oxalate ion ($C_2O_4^{2-}$) in the juice was estimated quantitatively by gravimetric method [17]. The juice was filtered through a Whatman-40 filter paper using vacuum pump. 50 mL of the filtrate was taken in a 200 mL beaker and acidified with acetic acid. The acidic solution was heated to boil and precipitated with boiling calcium chloride solution assisted by little ammonium chloride solution for effective precipitation. The solution was allowed to cool, treated with one third of its volume of 90% ethanol, and allowed to stand for 30 minutes. The precipitate was filtered through a pre-weighed sintered glass crucible and washed with warm water (50-60 °C) for several times until the chloride tested negative in the filtrate. The calcium oxalate was then washed with cold water, five times with ethanol, and several times with small volume of anhydrous diethyl ether. The precipitate was sucked dry at the pump for 10 minutes, the outside of the crucible was wiped dry with a clean cloth and kept in a vacuum desiccator for 10 minutes. The weight of the precipitate was recorded as $CaC_2O_4.H_2O$. Since this rapid method yields result of moderate accuracy, so the crucible with $CaC_2O_4.H_2O$ was heated at 475-525 °C in an electric muffle furnace for one hour. The crucible was allowed to cool in a desiccator and the precipitate was weighed as $CaCO_3$. This method is most satisfactory, since $CaCO_3$ is non-hygroscopic.

$$CaC_2O_4 \rightarrow CaCO_3 + CO$$

The quantity of oxalate ion ($C_2O_4^{2-}$) in 50 mL of banana pseudo-stem juice was calculated as:

$$\text{Total weight of } C_2O_4^{2-} \text{ in 50 mL juice} = \frac{Weight\,of\,CaCO_3 \times 88}{100.08}\,g$$

$$\text{Concentration of } C_2O_4^{2-} \text{ in the juice} = \frac{Weight\,of\,CaCO_3 \times 88 \times 20 \times 1000}{100.08}\,ppm$$

4.3. Estimation of Chloride Ion (Cl^-)

Chloride ion in the juice was estimated by the gravimetric method [17]. 50 mL of juice in a 250 mL beaker was acidified (tested with litmus) with 0.1 M nitric acid followed by drop wise addition of silver nitrate solution (0.1 M) with constant stirring until the precipitation was completed and then a few drops in excess. Completion of precipitation was tested by allowing the precipitate to settle down and adding a few drops of $AgNO_3$ solution carefully to the supernatant solution. The mixture was heated at near boiling point with constant stirring until the whole precipitate coagulated and the supernatant liquid was clear. The mixture was then allowed to cool down to room temperature, and filtered through pre-weighted Gooch crucible. The precipitate was dried in a hot air oven for an hour at 130 °C to 160 °C. The crucible along with the precipitate was cooled to room temperature in a desiccator and weighed. From the difference of this weight and the weight of empty crucible, the weight of AgCl was known from which weight of Cl^- was calculated. The whole process of chloride estimation was carried out in subdued light as AgCl is sensitive to light.

The weight of chloride (Cl^-) in 50 mL juice

$$= \frac{Weight\ of\ ppt \times 35.45}{143.35}\,g$$

Chloride concentration in the juice $= Weight\,of\,ppt \times 0.24739 \times 20 \times 1000\,ppm$

4.4. Estimation of Nitrate Ion by Spectrometric Method

Nitrate ion (NO_3^-) in the juice was estimated using UV spectrometer (U-3210 spectrometer, Hitachi) [18]. Nitrate ion absorbs ultraviolet radiation at 220 nm but not at 275 nm. Organic matters, if present, absorb at both 220 nm and 275 nm. To make a distinction between the absorbance by nitrate and other organic matters, the sample's absorbance was recorded both at 275 nm as well as at 220 nm, and an empirical correction factor was applied to the 220 nm measurements to distinguish the nitrate absorbance from that of organic matters.

Procedure: The juice sample was diluted to ten times of its original volume and acidified with 1N HCl to prevent interference due to the absorptions by OH⁻ and CO_3^{2-}, both of which can absorb at 220 nm. Hydrochloric acid is used because Cl⁻ does not absorb light in the 250–290 nm region of the spectrum. The absorbance readings were recorded both at 220 nm and 275 nm using distilled water as a blank with the same amount of HCl.

Experimental absorbance due to NO_3^-

= Absorbance at 220 nm – 2 × absorbance at 275 nm

A standard curve was prepared in the range of 0.0 to 50 mg/L of NO_3^- at intervals of 10 following the same procedure as described for the sample. The concentration of nitrate ion NO_3^- was calculated from the following relation which is derived from the calibration curve (Fig. 1).

$$x = \frac{y - 0.9295}{0.1021} \text{ ppm}$$

Nitrate ion (NO_3^-) concentration (in ppm) in the original juice sample is calculated as:

Nitrate ion (NO3⁻) in juice

$$= \frac{y - 0.9295 \times 10}{0.1021} \text{ ppm}$$

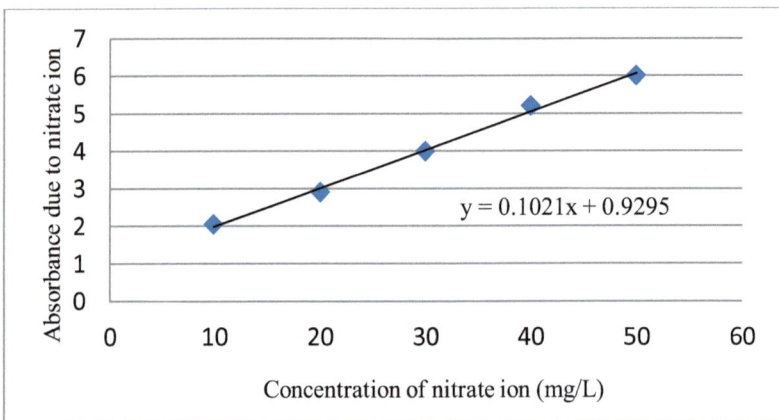

Fig. (1). Calibration curve (NO_3^- concentration *vs.* absorbance).

4.5. Estimation of Phosphate by Spectrometric Method

The phosphate content in juice samples was estimated by UV-Visible spectrometric method as per the reported procedure [19]. For this experiment, the required reagents were prepared as follows:

a. Ammonium molybdate solution:
   i. Dissolved 5.0 g of ammonium molybdate in 35 mL of distilled water.
   ii. Added 56 mL of conc. H_2SO_4 to 80 mL distilled water and cooled.

The two solutions were mixed and the volume was made up to 200 mL with distilled water and cooled.

b. Stannous chloride solution: 0.5 g of stannous chloride was dissolved in 20 mL glycerol by heating in a water bath.
c. Standard phosphate solution: 0.2194 g of pre-dried anhydrous potassium hydrogen phosphate, K_2HPO_4 was dissolved in distilled water and the volume was made up to 100 mL. The solution was then diluted 100 times (10 mL to 1000 mL). This was used as the standard phosphate solution containing 10 mg P/L (1 mL $\equiv$ 0.01 mg P).

Procedure: 50 mL of juice sample was digested in conc. HNO_3, evaporated to near dryness, 40 mL distilled water added and filtered. The volume of the filtrate was made up to 50 mL with distilled water. This solution was used as the stock for the experiment.

For UV-Vis measurement, 10 mL of the stock solution was diluted to 5 times its volume by adding 40 mL distilled water. To this 50 mL solution was added 2 mL of ammonium molybdate solution and 5 drops of stannous chloride solution. A blue colored solution appeared. It confirmed the presence of phosphate. After 5 minutes the optical density of the sample was recorded at 690 nm in a UV-Vis spectroscope using distilled water as a blank with the same amount of the chemicals. The concentration of phosphate was estimated with the help of a standard curve which was prepared in the range of 0.0 to 5.0 mg/L of PO_4^{3-} at intervals of 0.5 following the same procedure as described for the sample.

Phosphate (PO^{3-}_4) concentration in the juice sample was calculated using the following relation which was derived from the calibration curve (Fig. **2**).

$$x = \frac{y - 0.0777}{0.0331} \, ppm$$

Phosphate ion (PO_4^{3-}) in the original juice solution (in ppm) was calculated as:

$$\text{Phosphate ion (PO}_4^{3-}) \text{ concentration in the juice } = \frac{y - 0.0777 \times 5}{0.0331} \, ppm$$

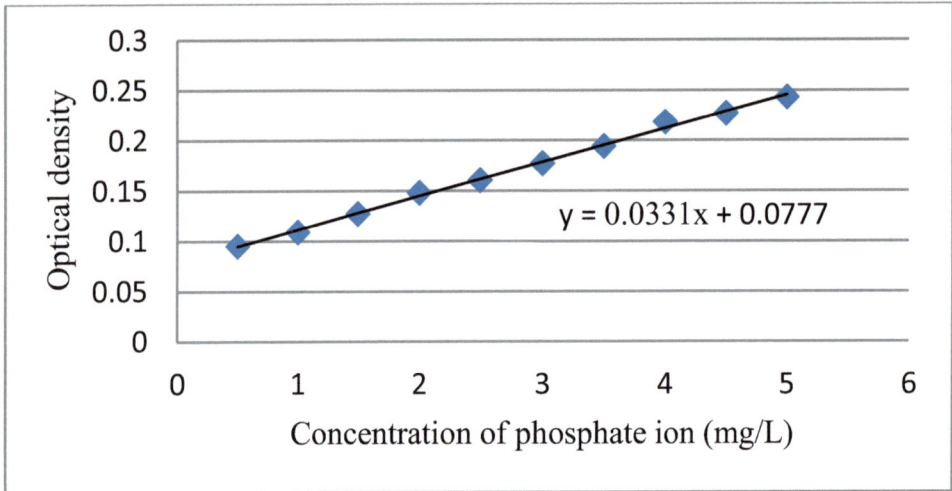

Fig. (2). Calibration curve (PO_4^{3-} concentration *vs.* optical density).

4.6. Estimation of Sodium (Na^+) and Potassium (K^+) by Flame Photometry

For the estimation of Na^+ and K^+ ions by Flame Photometry, the juice was first subjected to conc. nitric acid digestion as was done in case of phosphate ion estimation (Section 4.5) [20]. Estimation was carried out using Systronics, Flame Photometer 128. The sample for Flame Photometry was prepared by diluting 100 times its original volume by taking 5 mL of the sample solution (from nitric acid digestion) in a 500 mL volumetric flask and filling up to the mark with redistilled water. The standard solutions were prepared as follows:

100 ppm solution of Na^+ was prepared by dissolving 0.0254 g of NaCl (GR, 99.5%) in 100 mL of distilled water in a volumetric flask. Similarly, 100 mL of 100 ppm K^+ solution was prepared by dissolving 0.0191 g of KCl (GR 99.5%) in 100 mL distilled water. Four other standard solutions *viz.* 40 ppm, 20 ppm, 10 ppm and 5 ppm were prepared from 100 ppm standard solution by dilution technique.

$$\text{Na}^+ \text{ and K}^+ \text{ in the original juice } = \textit{Flame Photometer reading} \times 100 \; ppm$$

4.7. Estimation of Trace Metals in Juice by Atomic Absorption Spectroscopy

The presence of a few trace metals in the juice was estimated by Atomic Absorption Spectroscopy (Varian Spectra AA-220). The samples for Atomic Absorption Spectroscopy were prepared as follows [20]:

50 mL of juice was made strongly acidic with 6 mL conc. HNO_3 and evaporated to near dryness. It was then cooled to room temperature, dissolved in distilled water and filtered. The filtrate was transferred to 50 mL volumetric flask and filled with distilled water up to the mark. The solution so prepared was taken for the estimation of trace metals by Atomic Absorption Spectroscopy.

5. METHODS OF ORGANIC ANALYSIS

5.1. Preparation of Ethyl Acetate Extract (crude)

200 mL banana leaf sheath juice was mixed with 100 mL ethyl acetate in a separating funnel, shaken well and the organic layer separated. This was repeated twice with 50 mL ethyl acetate each. Combined organic phase was dried over anhydride sodium sulphate for two hours. After filtration, the solvent was removed in a rotary vacuum evaporator. Traces of solvents were further removed under high vacuum and the residue weighed.

5.2. Isolation of Organic Components from the Crude Extract

TLC (in iodine chamber) on the ethyl acetate crude extract indicated the presence of at least five compounds out of which two were major. These two compounds were isolated by column chromatography over silica gel (60-120 mesh) using 15% ethyl acetate in petroleum ether.

5.3. General Experimental Procedures for the Identification of the Isolated Compounds

The isolated compounds were characterized by melting point and spectral study. Melting points were recorded in a melting point apparatus (Scientific Device, India, type MP-D in open capillary) and uncorrected. IR spectra were recorded with a FT-IR spetrometer (Model Perkin Elmer spectrum RX I FT-IR system) on KBr pallets. All ^1H NMR spectra were recorded on Bruker-300 MHz spectrometer in $CDCl_3$ as the solvent using TMS as the internal reference. All GC-MS spectra were recorded on Perkin-Elmer-Clarus 600 Spectrometer with Elite 5 MS column, dimension 30m × 250µm. The injection temperature was fixed at 280 °C. The oven temperature was initially held at 250 °C for 2 minutes, increased to 270 °C at 2 °C/ min and held at 270 °C for 2 min. Helium was used as the carrier gas.

6. RESULTS AND DISCUSSION

6.1. Inorganic Analysis

Banana pseudo-stem is highly rich in water which accounts for about 95% of mass. A sample of fresh pseudo-stem weighing 25.00 kg yielded 17.26 liters of juice along with soluble phyto-chemicals and 0.831 kg of dry fiber. Analysis of the juice shows that all the soluble inorganic constituents together account for about 0.4%, total soluble organic compounds account for about 0.7% and total fiber accounts for about 3.3%. The results are shown in Tables **1** - **5**.

Table 1. Quantity of juice and fiber from pseudo-stem and tender core (floral stalk).

Entry	Material	Fresh Weight (kg)	Volume of Juice (L)	pH	Weight of Dry Fiber (kg)
1	Leaf sheath	25.00	17.26	6.9	0.831
2	Tender core	9.90	5.30	6.9	0.340

Table 2. Results of chemical and spectroscopic investigation of leaf sheath juice.

Entry	Ion	Concentration (ppm)	Methods of Determination
1	Na^+	619.50	Flame photometry
2	K+	2250.00	Flame photometry
3	Cl^-	549.60	Gravimetric analysis using silver nitrate
4	NO_3^-	136.00	UV spectrometry
5	$C_2O_4^{2-}$	1820.30	Gravimetric using calcium chloride
6	PO_4^{3-}	25.00	UV spectrometry

Table 3. Results of estimation of trace metals in the juice of leaf sheath by Atomic Absorption Spectroscopy.

Entry	Metal	Concentration (ppm)
1	Al	1.599
2	Ca	4.883
3	Cd	0.016
4	Co	0.011
5	Cr	0.009

(Table 3) cont.....

Entry	Metal	Concentration (ppm)
6	Cu	0.079
7	Fe	5.536
8	Mg	1.450
9	Mn	0.205
10	Ni	0.010
11	Pb	0.032
12	V	3.349
13	Zn	1.995

Chemical investigations on the two samples of leaf sheath juice and tender core juice reveal that oxalate ($C_2O_4^{2-}$) and chloride (Cl^-) are the two major anionic radicals present (Tables **2** and **4**). Oxalate concentration is higher than that of chloride in both the samples. Higher quantity of oxalate is found in the juice of leaf sheath (1820.3 ppm) as compared to that in tender core juice (1539.8 ppm), but both samples have almost equal quantities of chloride ion, 549.60 ppm and 578.89 in leaf sheath juice and tender core juice, respectively. Oxalate concentration is 2.5 to 3 times higher than that of chloride. Juice samples also contain nitrate and phosphate, which are present in lesser quantity as compared to the amounts of chloride and oxalate ions. In case of phosphate ion, it is almost same in both the samples, but its quantity (about 25 ppm) is least among all the anionic radicals. The amount of nitrate ion (NO_3^-) in the leaf sheath juice (136 ppm) is almost twelve times of that in the tender core juice (11.8 ppm).

Table 4. Results of chemical and spectroscopic investigation of tender stem juice.

Entry	Ion	ppm	Methods of Determination
1	Na^+	642.00	Flame photometry
2	K^+	2350.00	Flame photometry
3	Cl^-	578.89	Gravimetric using silver nitrate
4	NO_3^-	11.80	UV spectrometry
5	$C_2O_4^{2-}$	1539.80	Gravimetric using calcium chloride
6	PO_4^{3-}	22.00	UV spectrometry

Table 5. Results of estimation of trace metals in the juice of tender stem by Atomic Absorption Spectroscopy.

Entry	Metal	ppm
1	Al	0.714
2	Ca	1.994
3	Cd	0.002
4	Co	0.012
5	Cr	0.015
6	Cu	0.078
7	Fe	4.536
8	Mg	0.011
9	Mn	0.079
10	Ni	ND
11	Pb	0.042
12	V	0.983
13	Zn	2.391

Potassium is the major cationic constituent in both the juice samples, also not widely different in concentration. While potassium is 1820.30 ppm in leaf sheath juice, it is 1539.80 ppm in tender core juice. The second major cation in both the juices is sodium which is present almost in equal quantities, 619.50 ppm and 642.00 ppm in leaf sheath juice and tender core juice, respectively. Potassium concentration is 2 to 3 times of that of sodium concentration. Presence of several other metals in trace amounts is recorded in both the juice samples. Among the trace metals in the leaf sheath juice, Fe, Ca and V are the majors followed by Zn, Al and Mg (Table 3). In the tender core juice, Fe is the major among the trace metals followed by Zn, Ca and V (Table 5). Other trace metals detected in the both the juice samples are Co, Ni, Cr, Cu, Mn, Pb and Cd.

6.2. Organic Analysis

Extraction of 200 mL leaf sheath juice with ethyl acetate yielded a brown colour semi-solid which displayed the presence of at least five compounds on TLC (in iodine chamber). Two of the compounds were major which were isolated by column chromatography, and these are coded as Leave Sheath Compound LSC-1 and LSC-2. Results are summarized below:

Weight of crude extract (brown semi-solid) = 2.0018 g
Weight of crude extract used for column separation = 0.5394 g
Weight of the component, coded LSC-1 = 0.239 g
Weight of the component, coded LSC-2 = 0.146 g

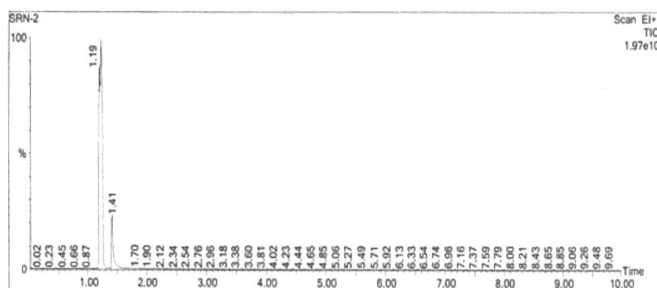

Fig. (3). GC of LSC-1.

GC of LSC-1 (Fig. **3**) confirms single compound purity. The ¹H NMR spectrum (Fig. **4**) indicates the presence of two methyl groups – one at σ 2.36 (indicating acetonic methyl) and the other at σ 3.84 (indicating methoxy group). Presence of one pair of protons at δ 6.6 and δ 7.45 with *J* value of 16.2 Hz suggests the presence of a *trans* olefinic bond in the molecule. Presence of four aromatic protons in two sets, each with two equivalent protons is an indication of a 1,4-disubstituted benzene ring in the molecule. These spectral evidences suggest that the compound LSC-1 may have a structure of (*E*)-4-(4-methoxyphenyl)but-3-en-2-one (Table **6**). This structure for LSC-1 is further supported by ¹³C NMR (Fig. **5**), IR spectrum (Fig. **6**) and Mass spectral data (Fig. **7**).

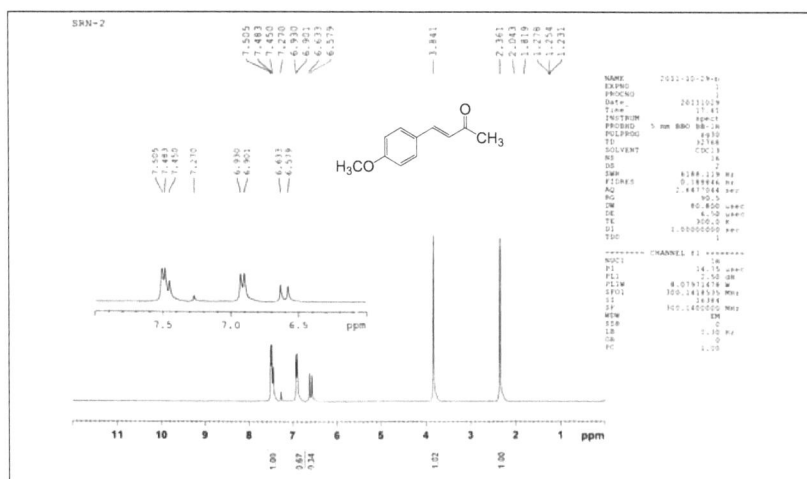

Fig. (4). ¹H NMR spectrum of the compound LSC-1.

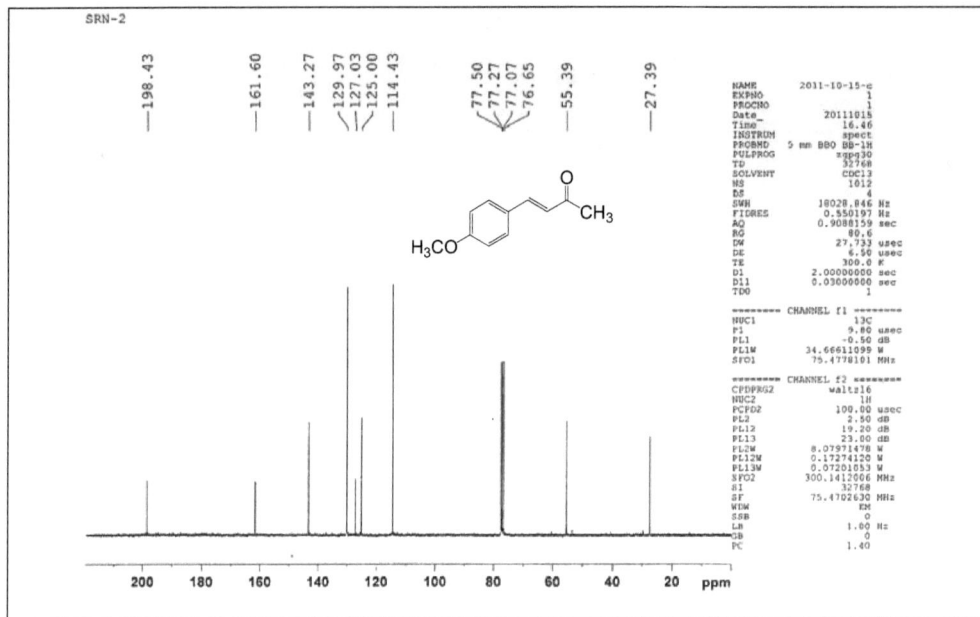

Fig. (5). ^{13}C NMR spectrum of the compound LSC-1.

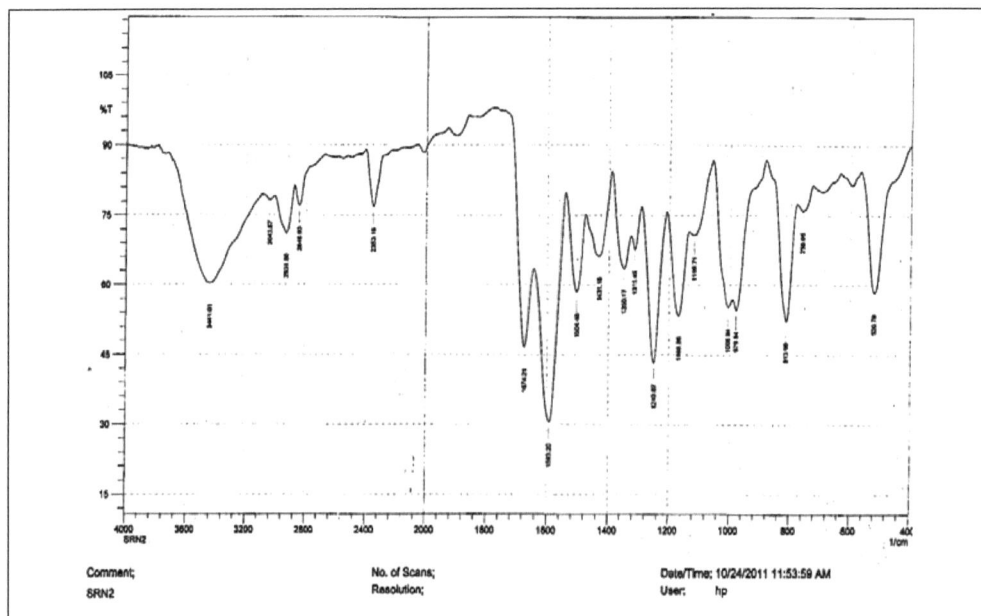

Fig. (6). IR spectrum of the compound LSC-1.

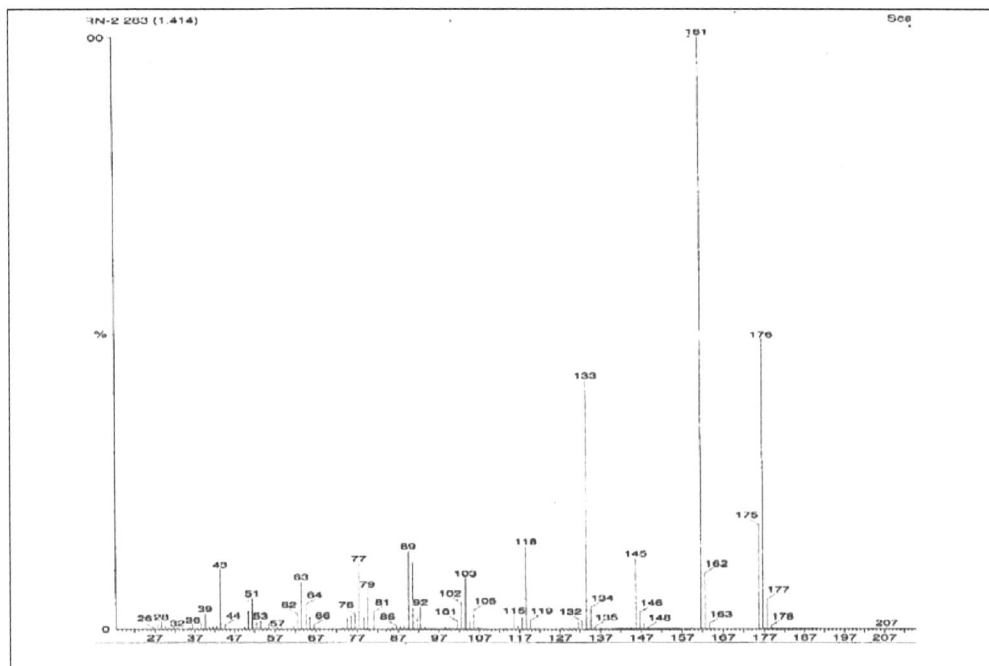

Fig. (7). MS of the compound LSC-1.

Table 6. Isolated organic compounds from leaf sheath juice.

Code	Name	Structure	Amount in Crude Extract (%)	Amount in Juice (% wt/v)
LSC-1	*(E)-4-(4-Methoxyphenyl)but-3-en-2-one*		*44.34*	*0.443*
LSC-2	*(1E,4E)-1,5-Bis(4-methoxyphenyl)penta-1,4-dien-3-one*		*27.1*	*0.271*

Spectral data of compound LSC-1

State: Yellow solid

Melting point: 72 °C

IR, v cm^{-1}: 521, 760, 814, 981, 1007, 1119, 1169, 250, 1315, 1350, 1431, 1504,

1593, 1674, 2353, 2847, 2932, 3044

^{1}H NMR (300 MHz, CDCl$_3$) δ (ppm): 2.36 (s, 3H, Me), 3.84 (s, 3H, ArOMe), 6.60 (d, J = 16.2 Hz, 1H, olefinic), 6.92 (d, J =8.7 Hz, 2H, aromatic), 7.45-7.50 (m, 2H, aromatic, 1H, olefinic).

^{13}C NMR (75 MHz, CDCl$_3$) δ (ppm): 27.39, 55.39, 114.43, 125.00, 127.03, 129.97, 143.27, 161.60, 198.43.

MS, m/z: 176 (M$^+$, ~50%), 161(100%), 145(15%), 133(45%), 118(20%), 89(20%), 77(15%), 63(10%), 43(10%).

GC of LSC-2 (Fig. **8**) confirms single compound purity. The ^{1}H NMR spectrum (Fig. **9**) indicates the presence of six equivalent protons at σ 3.85 (possibility of two methoxy groups). The pattern of the spectrum in the region σ 6.9 to 7.7 indicates the presence of two sets of aromatic protons and a doublet with J=15.9 Hz, indicating the presence of *trans* olefinic protons. Considering the molecular ion peak at 294.1 (Fig. **10**) with 100% intensity along with proton NMR, the most probable structure one would suggest for LSC-2 is (1E,4E)-1,5-bis-4-methoxyphenyl)penta-1,4-dien-3-one (Table **6**). This structure for LSC-2 is supported by other spectral evidences *viz.* ^{13}C NMR (Fig. **11**) and IR (Fig. **12**) spectra.

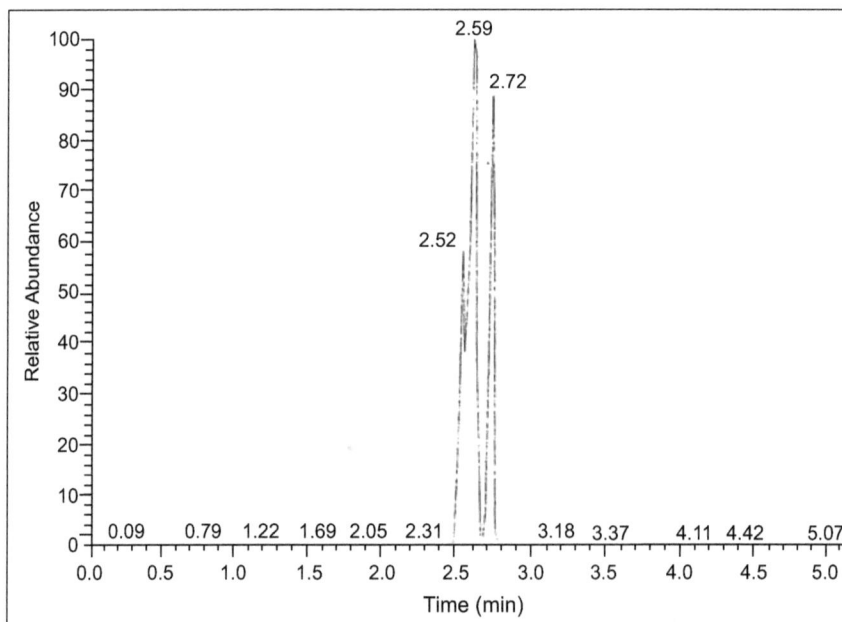

Fig. (8). GC of LSC-2.

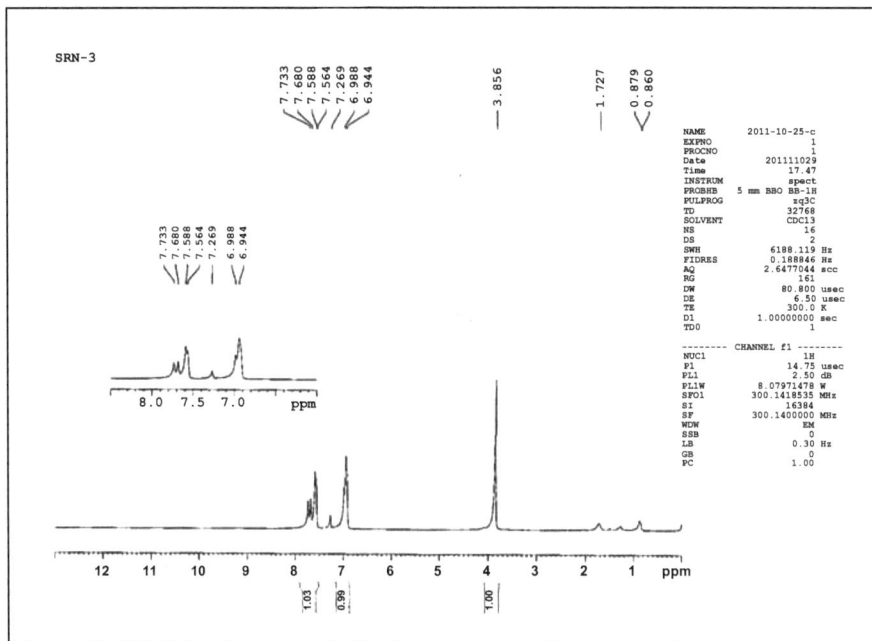

Fig. (9). ¹H NMR spectrum of the compound LSC-2.

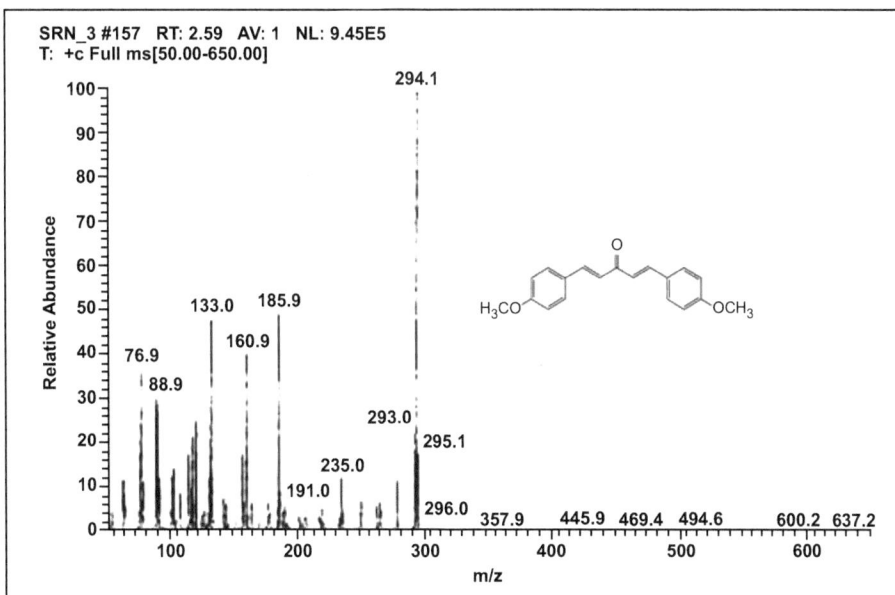

Fig. (10). Mass spectrum of LSC-2.

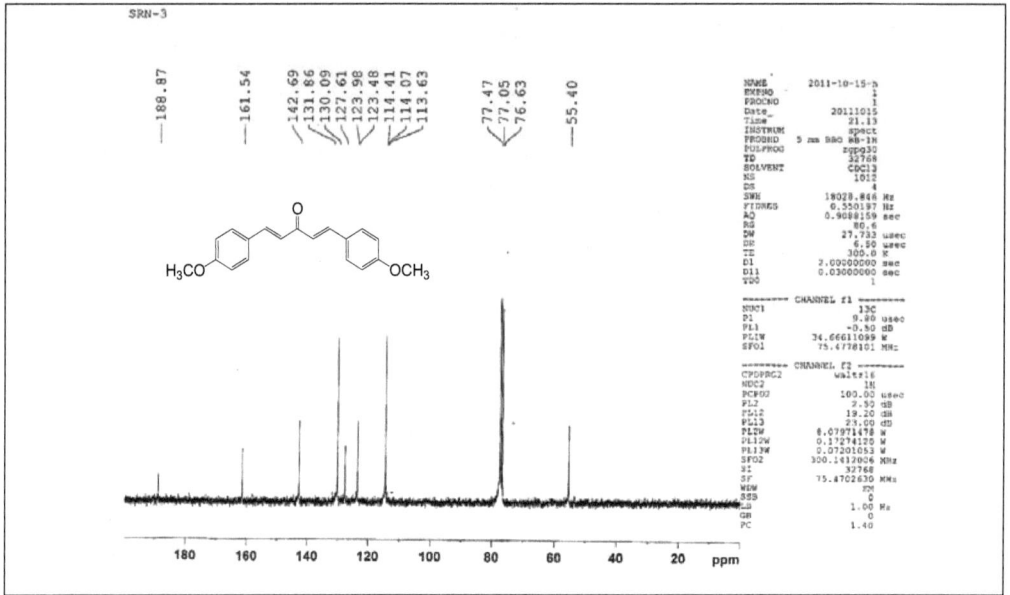

Fig. (11). ^{13}C NMR spectrum of the compound LSC-2.

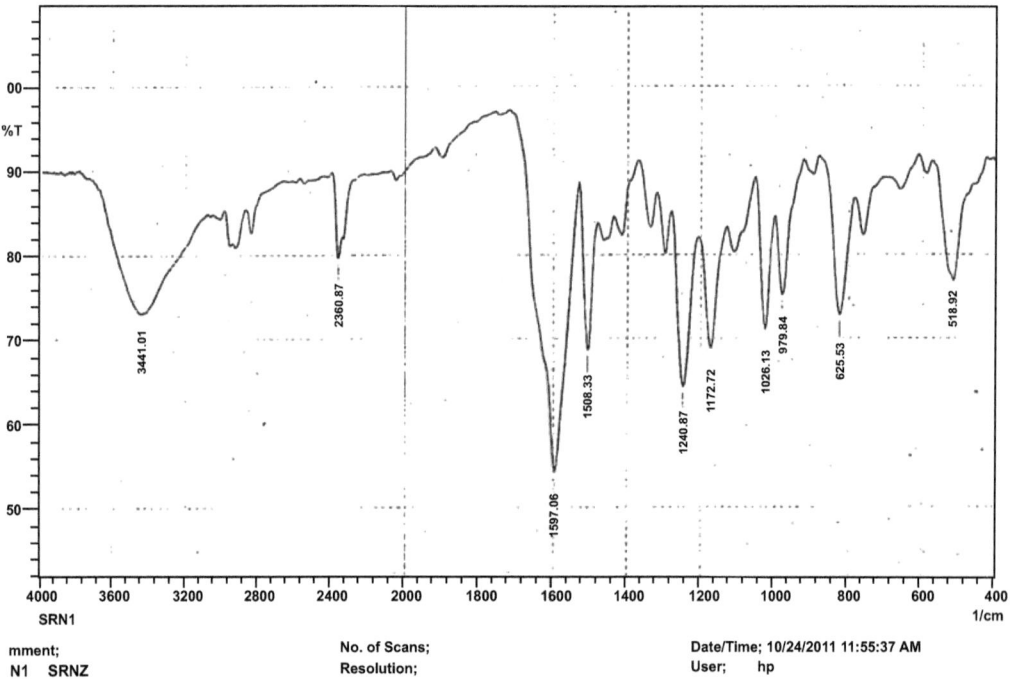

Fig. (12). IR spectrum of the compound LSC-2.

Spectral data of compound LSC-2

State: Light yellow crystalline solid
Melting point: 128 °C
IR, v cm^{-1}: 517, 826, 980, 1026, 1173, 1250, 1508, 1597, 1678, 2361.

^{1}H NMR (300 MHz, CDCl$_3$) δ(ppm): 3.86 (s, 6H, ArOMe), 6.94 – 6.99 [m, 6H (2H-olefinic, 4H-aromatic)], 7.58 (d, *J* = 7.8 Hz, 4H, aromatic), 7.71 (d, *J* = 15.9, 2H, olefinic)

^{13}C NMR (75 MHz, CDCl$_3$) δ(ppm): 55.40, 114.07, 123.98, 127.61, 130.09, 142.69, 161.54, 188.87

MS, m/z: 294.1(M$^+$, ~100%), 185.9(50%), 160.9(40%), 133(50%), 88.9(30%), 76.9(36%)

LSC-1 [(*E*)-4-(4-methoxyphenyl)but-3-en-2-one] and LSC-2 [(1*E*,4*E*)-1,5-bis-4-methoxyphenyl)penta-1,4-dien-3-one] are so far not reported in any natural source. Therefore, banana plant pseudo-stem appears to be the first ever reported natural source for these two compounds [2]. Both the compounds belong to the class of α,β-unsaturated acyclic ketones which find wide applications as useful key reagents in organic synthesis. Their use as substrates in synthetically important reactions includes Michael addition [21], cycloaddition [22], Morita Baylis-Hillman reaction [23], *etc.* The most common access to (*E*)-α,β-unsaturated ketones is by the Claisen-Schmidt condensation of aldehydes and ketones under basic conditions [24]. Preparation and structures of LSC-1 and LSC-2 have been reported [25, 26].

Compound (*E*)-4-(4-methoxyphenyl)but-3-en-2-one is an analogue to curcumin. Curcumin is a yellow compound isolated from the rhizome of the herb *Curcuma longa* L., which possesses multifunctional pharmacological properties including apoptosis in a variety of tumor cells [27 - 29]. Although curcumin is remarkably non-toxic and has promising anti-inflammation and anti-cancer activities, its poor bioavailability and pharmacokinetic profiles have limited its application in anti-cancer therapies [29, 30]. During the last decade, synthetic modifications of curcumin, which were aimed at enhancing its bioactivities, have been intensively studied. One of such modified synthetic compound is (1*E*,4*E*)-1,5-bis-4-methoxyphenyl)penta-1,4-dien-3-one and can be synthesized by reacting 4-methoxybenzaldehyde with acetone in presence of sodium hydroxide [25]. However,ss it is reported to have less anti-tumor and chemo-preventive activity than curcumin [29]. Studies on antioxidant and *in vitro* anti-tumor activity of compound (*E*)-4-(4-methoxyphenyl)but-3-en-2-one have also been reported [25, 31].

7. CONCLUSION

Banana plant pseudo-stem consists of nearly 95% juice and less than 5% fiber. The juice is quite rich in potassium and oxalate, moderately rich in sodium and chloride. In addition, nitrate, phosphate and a couple of heavy metals in trace concentrations are present. Two important bioactive organic molecules, namely (*E*)-4-(4-methoxyphenyl)but-3-en-2-one and (1*E*,4*E*)-1,5-bis(4-methoxyphenyl) penta-1,4-dien-3-one, have been isolated from the juice, and characterized.

REFERENCES

[1]   El-Sayed, M.; Mansour, O.Y.; Selim, I.Z.; Ibrahim, M.M. Identification and utilization of banana plant juice and its pulping liquor as anti-corrosive materials. *J.Sci. Ind. Res.,* **2001**, *60*, 738-747.

[2]   Neog, S.R. *Banana Plant: A Renewable Source of Potassium Chloride and Potassium Carbonate (PhD Thesis),* **2014**,

[3]   Jones, V.; Warjri, L. Medicinal properties of the banana plant. *Med India,* **2019**. https://www.medindia.net/patients/lifestyleandwellness/medicinal -properties-of-the-banana-plant.htm

[4]   Tyler, V.M.; Premila, M.S. *A Clinical Guide to the Healing Plants of Traditional Indian Medicine,* 1st ed; Routledge: New Delhi, **2006**.

[5]   *Banana stem juice for detox and weight loss,* **2018**, www.timesnownews.com /health/banana-ste--juice-for-detox-and-weight-loss

[6]   Reddy, A.J.; Dubey, A.K.; Handu, S.S.; Sharma, P.; Mediratta, P.K.; Ahmed, Q.M.; Jain, S. Anticonvulsant and antioxidant effects of *Musa sapientum* stem extraction: Acute and chronic experimental models of epilepsy. *Pharmacognosy Res.,* **2018**, *10*(1), 49-54. [PMID: 29568187]

[7]   Mallick, C.; Mandal, S.; Barik, B.; Bhattacharya, A.; Ghosh, D. Protection of testicular dysfunctions by MTEC, a formulated herbal drug, in streptozotocin induced diabetic rat. *Biol. Pharm. Bull.,* **2007**, *30*(1), 84-90. [http://dx.doi.org/10.1248/bpb.30.84] [PMID: 17202665]

[8]   Mohapatra, D.; Mishra, S.; Sutar, N. Banana and its by-product utilisation: An overview. *J. Sci. Ind. Res. (India),* **2010**, *69*, 323-329.

[9]   Saravanan, K.; Aradhya, S.M. Polyphenols of pseudostem of different banana cultivars and their antioxidant activities. *J. Agric. Food Chem.,* **2011**, *59*(8), 3613-3623. [http://dx.doi.org/10.1021/jf103835z] [PMID: 21405133]

[10]   Halliwell, B.; Aeschbach, R.; Löliger, J.; Aruoma, O.I. The characterization of antioxidants. *Food Chem. Toxicol.,* **1995**, *33*(7), 601-617. [http://dx.doi.org/10.1016/0278-6915(95)00024-V] [PMID: 7628797]

[11]   Branen, A.L. Toxicology and biochemistry of butylated hydroxyanisole and butylated hydroxytoluene. *J. Am. Oil Chem. Soc.,* **1975**, *52*(2), 59-63. [http://dx.doi.org/10.1007/BF02901825] [PMID: 805808]

[12]   Velioglu, Y.S.; Mazza, G.; Gao, L.; Oomah, B.D. Antioxidant activity and total phenolics in selected fruits, vegetables, and grain products. *J. Agric. Food Chem.,* **1998**, *46*, 4113-4117. [http://dx.doi.org/10.1021/jf9801973]

[13]   Saravanan, K.; Aradhya, S.M. Potential nutraceutical food beverage with antioxidant properties from banana plant bio-waste (pseudostem and rhizome). *Food Funct.,* **2011**, *2*(10), 603-610. [http://dx.doi.org/10.1039/c1fo10071h] [PMID: 21915417]

[14] Espín, J.C.; García-Conesa, M.T.; Tomás-Barberán, F.A. Nutraceuticals: facts and fiction. *Phytochemistry,* **2007**, *68*(22-24), 2986-3008. [http://dx.doi.org/10.1016/j.phytochem.2007.09.014] [PMID: 17976666]

[15] Lee, S.K.; Ng, L.L.; Lee, S.I. Experiments with banana trunk juice as a neuromuscular blocker. *Aust. J. Exp. Biol. Med. Sci.,* **1980**, *58*(6), 591-594. [http://dx.doi.org/10.1038/icb.1980.60] [PMID: 7271597]

[16] Alwi, H.; Idris, J.; Musa, M.; Hamid, K.H.K. A preliminary study of banana stem juice as a plant-based coagulant for treatment of spent coolant wastewater. *J. Chem., Article ID,* **2013**, *165057*, 1-7. [http://dx.doi.org/10.1155/2013/165057]

[17] Basset, J.; Denney, J.R.C.; Jeffery, G.H.; Mendham, J., Eds. *Vogel's Textbook of Quantitative Inorganic Analysis,* 4th ed; English Language Book Society/Longman, **1986**.

[18] APHA Method 4500-NO$_3$. In: *Standard Methods for the Examination of Water and Wastewater,* 18th ed; American Public Health Association & American Water Works Association: New York, **1992**.

[19] Trivedy, R.K.; Goel, P.K.; Trisal, C.L. *Practical Methods in Ecology and Environmental Science*; Environmental Publications: Karad, India, **1987**.

[20] Kour, H. *Spectroscopy,* 1st ed; Pragatiprakashan: New Delhi, **2017**.

[21] a) Krause, N.; Hoffmann-Röder, A. Recent advances in catalytic enantioselective Michael additions. *Synthesis,* **2001**, (2), 171-196. [http://dx.doi.org/10.1055/s-2001-10803] b) Betancort, J.M.; Barbas, C.F., III Catalytic direct asymmetric Michael reactions: taming naked aldehyde donors. *Org. Lett.,* **2001**, *3*(23), 3737-3740. [http://dx.doi.org/10.1021/ol0167006] [PMID: 11700126] c) Almasi, D.; Alonso, D.A.; Nájera, C. Organocatalytic asymmetric conjugate additions. *Tetrahedron Asymmetry,* **2007**, *18*, 299-365. [http://dx.doi.org/10.1016/j.tetasy.2007.01.023]

[22] a) Barluenga, J.; Fanlo, H.; López, S.; Flórez, J. [4+1]/[2+1] Cycloaddition reactions of Fischer carbene complexes with α,β-unsaturated ketones and aldehydes. *Angew. Chem. Int. Ed. Engl.,* **2007**, *46*(22), 4136-4140. [http://dx.doi.org/10.1002/anie.200605167] [PMID: 17440911] b) Hernández-Toribio, J.; Gómez Arrayás, R.; Martín-Matute, B.; Carretero, J.C. Catalytic asymmetric 1,3-dipolar cycloaddition of azomethine ylides with α,β-unsaturated ketones. *Org. Lett.,* **2009**, *11*(2), 393-396. [http://dx.doi.org/10.1021/ol802664m] [PMID: 19093844]

[23] Basavaiah, D.; Rao, A.J.; Satyanarayana, T. Recent advances in the Baylis-Hillman reaction and applications. *Chem. Rev.,* **2003**, *103*(3), 811-892. [http://dx.doi.org/10.1021/cr010043d] [PMID: 12630854]

[24] a) Wang, Z. Claisen-Schmidt condensation. In: *Comprehensive Organic Name Reactions and Reagents*; John Wiley & Sons, **2010**. [http://dx.doi.org/10.1002/9780470638859.conrr145] b) Carey, F.A. *Organic Chemistry,* 3rd ed; McGraw-Hill: New York, **1996**.

[25] Handayani, S.; Arty, I.S. Synthesis of hydroxyl radical scavengers from benzalacetone and its derivatives. *J. Physiol. Sci.,* **2008**, *19*(2), 61-68.

[26] Sambyal, A.; Kour, M.; Anthal, S.; Bamzai, R.K.; Kant, R.; Gupta, V.K. (*E*)-4-(4-Methoxyphenyl)-ut-3-en-2-one. *ActaCryst. Struct. Rep.,* **2012**, *E68*, o1183.

[27] Jagetia, G.C.; Aggarwal, B.B. "Spicing up" of the immune system by curcumin. *J. Clin. Immunol.,* **2007**, *27*(1), 19-35. [http://dx.doi.org/10.1007/s10875-006-9066-7] [PMID: 17211725]

[28] a) Aggarwal, B.B.; Kumar, A.; Bharti, A.C. Anticancer potential of curcumin: preclinical and clinical studies. *Anticancer Res.,* **2003**, *23*(1A), 363-398. [PMID: 12680238] b) Mukherjee (nee Chakraborty), S.; Ghosh, U.; Bhattacharyya, N.P.; Bhattacharya, R.K.; Dey, S.; Roy, M. Curcumin-induced apoptosis in human leukemia cell HL-60 is

associated with inhibitionof telomerase activity. *Mol. Cell. Biochem.,* **2007**, *297*, 31-39.
[http://dx.doi.org/10.1007/s11010-006-9319-z]

[29]   Hsu, C-H.; Cheng, A-L. Clinical studies with curcumin. *Adv. Exp. Med. Biol.,* **2007**, *595*, 471-480.
[http://dx.doi.org/10.1007/978-0-387-46401-5_21] [PMID: 17569225]

[30]   Garcea, G.; Jones, D.J.L.; Singh, R.; Dennison, A.R.; Farmer, P.B.; Sharma, R.A.; Steward, W.P.;
Gescher, A.J.; Berry, D.P. Detection of curcumin and its metabolites in hepatic tissue and portal blood
of patients following oral administration. *Br. J. Cancer,* **2004**, *90*(5), 1011-1015.
[http://dx.doi.org/10.1038/sj.bjc.6601623] [PMID: 14997198]

[31]   Motohashi, N.; Yamagami, C.; Tokuda, H.; Okuda, Y.; Ichiishi, E.; Mukainaka, T.; Nishino, H.; Saito,
Y. Structure-activity relationship in potentially anti-tumor promoting benzalacetone derivatives, as
assayed by the epstein-barr virus early antigen activation. *Mutat. Res.,* **2000**, *464*(2), 247-254.
[http://dx.doi.org/10.1016/S1383-5718(99)00198-9] [PMID: 10648911]

CHAPTER 3

Use of Banana Plant Pseudo-stem Juice as the Substitute for Muriate of Potash in Agriculture: Application in the Cultivation of Rice

Abstract: Materials and methods for the use of banana plant pseudo-stem juice in the cultivation of rice paddy have been discussed. The yield of paddy from the soil treated with banana plant pseudo-stem juice is compared with that from the soil treated with muriate of potash. Juice treated soil shows a higher number of seeds per panicle, higher average seed weight as well as higher yield per unit of land as compared to potash treated soil. Overall yield in juice treated soil is 10% more than that in potash treated soil, and 40% more than that in soil not treated with potash. Thus, the banana plant pseudo-stem juice is a better replacement for potash fertilizer. Colour pictures of experimental plots are displayed.

Keywords: Organic fertilizer, Potash for rice cultivation, Substitute for MOP, Use of banana plant juice, Use of banana plant pseudo-stem.

1. INTRODUCTION

The threat of widespread hunger, especially in Asia, where the population depends mainly on crops like rice, wheat, *etc.*, practically remains unchanged even after the so-called green revolution in the 1960s. The negative consequences of green revolution technologies were due to extensive use of pesticides, mineral fertilizers and other agrochemical inputs. Agricultural policy all over the world seeks to promote technologically sound, economically viable, environmentally safe and socially acceptable use of natural resources such as land, water and biodiversity to promote sustainable development of agriculture, leading to an evergreen revolution.

The use of fertilizer in agriculture is a necessity to increase the production of food crops. However, to sustain the production of food crops for the longer future, soil fertility should not be compromised. Scientists have identified 16 essential nutrients, and according to the amount of requirement, these elements are grouped as primary nutrients, secondary nutrients and trace nutrients. Primary nutrients, also known as macronutrients, are those usually required in large amounts. These are carbon, hydrogen, nitrogen, oxygen, phosphorus, and potassium. The secondary nutrients such as calcium, magnesium, and sulfur are required in moderate amounts. Micro- or trace nutrients are required in tiny amounts compared to primary or secondary nutrients. Micronutrients are boron, chlorine, copper, iron, manganese, molybdenum, and zinc [1].

Fertilizer provides essential nutrients for plant or crop growth. Generally, soil provides all these nutrients to plants, crops or trees. But, in intensive cultivation of a particular plant (crop) at a particular place, the soil cannot provide all the essential nutrients to the intensively grown plants (crops) in proper proportion, in adequate quantities and at the appropriate time. To make the soil fertile, appropriate fertilizers should be used in proper proportions and adequate quantities. The fertilizer quantity to be applied to the soil can be calculated on the basis of the initial soil test reports. Nowadays, fertilizer adjustment or tailoring equations for different crops, different soils and climatic conditions are available [2].

Mined and synthesized inorganic fertilizers have significantly supported global population growth. It has been estimated that almost half the people on the Earth are currently fed as a result of synthetic nitrogen fertilizer use [3]. However, for good reasons, synthetic and inorganic fertilizers are nowadays less preferred, and organic fertilizers are being promoted. Organic fertilizers include naturally occurring organic materials such as vermicompost, chicken litters, worm castings, seaweeds, guanos, bone meals, *etc*. Origins of organic fertilizer are animals and plants. Non-organic chemical-based farming causes considerable degradation of soils. Therefore, alternative approaches have been tried out on the use of available and renewable resources of plant nutrients for complementing and supplementing the commercial fertilizers. Efforts are on to evaluate systematically the feasibility and efficiency of organic residues, not only for refurbishing the soil productivity but also promoting the efficiency of non-organic chemical fertilizers [4].

Organic fertilizers are known to improve biodiversity (soil life) and long-term productivity of soil [5], and also may prove a large depository for excess carbon dioxide [6]. Organic nutrients increase the abundance of soil organisms by providing organic matter and micronutrients for organisms such as fungal mycorrhiza [7] (which aid plants in absorbing nutrients) and can drastically reduce external inputs of pesticides, energy and fertilizer.

The combined use of organic and inorganic fertilizers in crop production has also been widely recognized as a way of increasing yield and improving the productivity of the soil [8]. Agricultural activities produce billions of tons of waste materials. With appropriate techniques, agricultural wastes can be recycled to produce natural fertilizer for crops. Recycling of agriculture residues should be an integral part of integrated plant nutrition, which is now being increasingly recognized as the strategy for sustaining high crop yield levels with minimal depletion of soil fertility or fall in its other quality aspects.

Potassium (K) is one of 16 nutrients that are essential for plant growth. Like nitrogen (N) and phosphorus (P), it is a macronutrient because plants require large amounts of K [9]. Potassium plays more roles in a plant than any other nutrient. It does not become a direct part of the plant structure, but it acts to regulate water balances, nutrient and sugar movement in plant tissue, drives starch and protein synthesis and legume nitrogen fixation. It improves yield, nutrient value, taste, color, texture and disease resistance of crops. Potassium is highly mobile within the plant. This implies that the crop may move K from old leaves to new plant parts [9]. Potassium is essential for photosynthesis and pod development in groundnut [10]. The potassium deficiency symptoms in the plant are (i) slow growth and weak stem, (ii) poorly developed root systems resulting in poor water use efficiency, (iii) poor N uptake, (iv) shriveled seed resulting in lower yield and (v) more susceptible to disease and winter kill.

Potash is available in different forms, but the commonly used forms in fertilizer are potassium chloride, potassium nitrate and potassium sulfate. It is the common name for various mined and manufactured salts that contain potassium in water-soluble form, the most common being potassium chloride. More than 60% of the potash produced by fertilizer industries is potassium chloride, which is better known as muriate of potash (MOP) in industries and commerce [11]. Most of the world reserves of potassium (K) were deposited as sea water in ancient inland oceans evaporated, and the potassium salts crystallized into beds of potash ore. These are the locations where potash is currently being mined today. The complete deposition required hundreds or even thousands of years [12]. As the result of the extensive use of potash as fertilizer, the reserves of high-grade potash will be quite limited with respect to global consumption and probable future demands. Considerable care is therefore required to protect existing resources against unwise use. Additional research to discover new potash deposits or to advance technical skills to discover and recover potash from low-grade sources would be helpful.

Banana plant contains good quantities of K, Na, Ca, Mg, Si and other essential elements, including the little amount of N and P [13]. Its pseudo-stem juice contains a considerable amount of potassium ion (2.25 g/L) along with chloride ion (0.31 g/L). pH of the juice is 6.90 just after the extraction from pseudo-stem

[14]. Banana is a common nutritious fruit and cultivated in many countries across the globe. It is a perennial plant and gives fruits only once. After harvesting a large volume of biomass goes waste. Therefore, exploration of the waste banana pseudo-stems for beneficial use will be significant and bring additional incentives to farmers. Rice paddy, wheat, mustard, chili and brinjal (eggplant) are a few common crops that are cultivated or can be cultivated in many parts of the world where the atmosphere is congenial for banana cultivation. This chapter of the book shows how banana plant pseudo-stem juice has been successfully used to supplement potash in the farming of rice paddy.

2. EXPERIMENTAL DETAILS

2.1. Materials

The experiment was conducted in Dhakuakhana, a locality in the district of Lakhimpur, Assam. For this experiment, post-harvest banana plant '*Musa balbisiana*' (popularly known as *Athiakol* or *Bhimkol* in Assamese) was selected. A few fresh pieces of post-harvest banana plant pseudo-stem were collected, and the juice was extracted from the pseudo-stem with the help of a sugar-cane juicer machine by squeezing several times. The volume of the fluid was measured, chemical analysis was done as per procedures described in Section 4, and the results are reported in Section 6 (Chapter 2). The juice was then used as the substitute for muriate of potash in the cultivation of *Oryza sativa* variety of rice (*Aijong*, kharif variety).

2.2. Details of Methodology

The crop cultivation was carried out in a randomized block design for four treatments. Before preparing the plots for cultivation, the soil was tested in the laboratory of the Department of Agriculture, Lakhimpur district, Assam and based on the test reports, quantities of fertilizers required for the cultivation of rice, *Oryza sativa* (*Aijong*, kharif variety) were recommended (Table **1**). Four plots of lands, each of size 4ft × 4ft, were used. Banana plant (*Musa -balbisiana*) juice as a source of potassium was applied to the soil of plot No. 2 (treatment T2) as a basal application of fertilizer, whereas commercial fertilizer, muriate of potash (MOP) was applied to plot No. 3 (treatment T3). No potash was applied in the plot No. 4 (treatment T4). Nitrogen in the form of urea and phosphorus in the form of single super phosphate (SSP) were applied in the plots 2, 3, and 4 (T2, T3 and T4) equally except plot No. 1 (treatment T1). Plot No.1 (treatment T1) was control. The different treatments in this experiment are shown in Table **2**.

Table 1. Soil test report.

Entry	Character	Value	Remark
1	pH	6.2	Slightly acidic
2	Organic carbon	0.12%	Low
3	Available phosphorous	44.5 kg/ha	Medium
4	Available potash	241kg/ha	Medium
5	Soil texture	SL (Silt loam)	-

Fertilizers recommended (in kg/bigha, 7.47 bigha = 1 hectare) for the cultivation of rice *Oryza sativa* (Aijong, kharif variety):
Farm Yard Manure (FYM) = 1500 kg
Urea = 14.2 kg
Single Super Phosphate (SSP)= 10.6 kg
Muriate of Potash (MOP) = 3.6 kg
Lime = not required

Table 2. Fertilizers applied in different plots for rice cultivation (g/plot).

Plot	Treatments	Urea (g)	SSP (g)	MOP (g)	Banana Plant Juice (Litres)
1	T1	Nil	Nil	Nil	Nil
2	T2	16	12	Nil	0.900
3	T3	16	12	4	Nil
4	T4	16	12	Nil	Nil

ProcEdure for Nursery Bed and Seedling Development: Requisite quantity of rice seeds (100 g of *Oryza sativa*) were cleaned and soaked in normal tap water. After 24 hours, water was completely drained out, and the seeds were packed in a piece of clean cotton cloth and kept hanging for the next 3 days. When the seeds started germination, these were uniformly broadcasted over the muddy surface of a nursery bed of 25 sq. ft. which was thoroughly puddled and leveled into a raised bed. During the preparation of the nursery bed, 6 kg FYM, 71 g SSP, 35 g urea and 19 g MOP were applied to the soil. The procedure was done in the 1st week of the month of July.

Procedure for Transplantation of Seedlings: 24 day-old paddy seedlings were lifted from the soil, roots rinsed in water to remove soil, leaves trimmed a few centimeters from the top, left on water overnight and next day transplanted. For transplantation of seedlings, a wet irrigated system was followed. The land used was a part of normal agricultural land, and transplantation was carried out towards the end of the month of July, the best suitable time for the variety. Four plots, each 4 ft x 4 ft, were prepared by repeated ploughing until muddy enough for transplantation. Dry FYM (1.7 kg per plot) was applied to the soil during the last ploughing. The full amount of SSP (12 g), MOP (4 g) and banana plant juice (0.9 L) but half of urea (8 g), as shown in Table **2**, were applied as basal at the time of final leveling of the plots. Seedlings of 25 days old were transplanted, maintaining 30 cm distance apart with 3-4 seedlings per hill (25 numbers of hill per plot). After 15 days of the plantation, the remaining half of urea (8 g) was dissolved in water and used to top-dress the plots in accordance with Table **2**. Before top dressing, water was drained out of the plots and allowed only after two days. Equal care was taken for all the treatments as per the advice of the Department of Agriculture. The plots were irrigated at regular intervals, and irrigation was stopped 10 days before harvesting.

Procedure for Collection of Seeds: When the seeds matured (120 days after transplantation for treatments T2, T3 & T4, 127 days for treatment T1), they were collected in such a way that the seeds could be compared as weight/seed, number of seed per panicle or pod and total production per plot. 10 panicles from each treatment were collected randomly and counted the numbers of seed in each panicle. Moreover, 300 seeds were randomly chosen from each treatment and weighed, and also weighed the total production in each treatment. Before weighing, seeds were cleaned and dried under the sun.

2.3. Statistical Analysis

'F Test' is one of the best methods to search any differences among more than two average values of different treatments. Therefore, with the help of 'F Test' attempt was made to search the differences among the average values of different treatments shown in Table **5**. All data collected were statistically analysed by analysis of variance (set up an ANOVA table). On the other hand, the procedure of searching for pair wise difference and multiple comparisons between two average values in the treatments, Tukey's method at 5% level, was applied.

3. RESULTS AND DISCUSSION

The quantities of macro and micronutrients present in banana plant pseudo-stem juice are shown in Tables **3** and **4** (for methods of estimation, please refer to Chapter 2).

Table 3. Major ions present in banana plant pseudo-stem juice.

Entry	Ions	ppm	g/L
1	Na^+	619.50	0.620
2	K^+	2250.00	2.250

(Table 3) cont.....

Entry	Ions	ppm	g/L
3	Cl^-	549.60	0.550
4	NO_3^-	136.00	0.136
5	$C_2O_4^{2-}$	1820.30	1.820
6	PO_4^{3-}	25.00	0.025

Table **3** reveals that the major cation found in banana plant pseudo-stem juice is potassium (2.250 g/L) and the major anion is oxalate (1.820 g/L). In addition,the juice also contains Cl^- (0.55 g/L), NO_3^- (0.136 g/L) and PO_4^{3-} (0.025 g/L) which are important nutrients for plant growth. Potassium, nitrate and phosphate are the usual components of commercial fertilizers widely used in rice cultivation. Potassium has several roles in plant growth and reproduction. It regulates the opening and closing of stomata during photosynthesis, and thus regulates the uptake of carbon dioxide. It activates enzymes and is essential for the production of Adenosine Triphosphate (ATP). It is required in large amounts and is considered second only to nitrogen when it comes to nutrients needed by plants. It affects the plant shape, size, color, taste and other measurements attributed to healthy produce. Plants absorb potassium in its ionic form, K^+ [15]. Nitrogen (N) is a part of all living cells and is a necessary part of all proteins, enzymes and metabolic processes involved in the synthesis and transfer of energy. It is also a part of chlorophyll, the green pigment of the plant that is responsible for photosynthesis. Nitrogen helps plants with rapid growth, increasing seeds and fruits production and improving the quality of leaf and forage crops [16]. Like nitrogen, phosphorus (P) is an essential part of the process of photosynthesis. It involves the formation of all oils, sugars, starches, *etc.*, and helps in the transformation of solar energy into chemical energy; proper plant maturation; withstanding stress. Phosphorus affects the rapid growth of plant and encourages blooming and root growth [16].

Banana plant pseudo-stem juice contains several micronutrients essential for plants, and among them, iron and calcium are the major (Table **4**).

Table 4. Micro nutrients in banana plant pseudo-stem juice.

Entry	Metal	ppm
1	Al	1.599
2	Ca	4.883
3	Cd	0.016
4	Co	0.011
5	Cr	0.009

(Table 4) cont.....

Entry	Metal	ppm
6	Cu	0.079
7	Fe	5.536
8	Mg	1.450
9	Mn	0.205
10	Ni	0.010
11	Pb	0.032
12	V	3.349
13	Zn	1.995

Iron is an essential micronutrient for plants. It plays a critical role in metabolic processes such as DNA synthesis, respiration, and photosynthesis. Further, many metabolic pathways are activated by iron, and it is a prosthetic group constituent of many enzymes. Iron imbalance in plants causes iron chlorosis [17]. Calcium is a secondary nutrient and required for various structural roles in the cell wall and membranes. It helps normal transport and retention of other elements. It is also believed to counteract the effect of alkali salts and organic acids within a plant [18]. Magnesium is also a secondary nutrient and required for the synthesis of chlorophyll in all green plants. It also helps to activate many plant enzymes needed for growth [19]. Copper is important for reproductive growth. It aids in root metabolism and helps in the utilization of proteins. Chloride (Cl) aids plant metabolism. Manganese (Mn) functions with enzyme systems involved in the breakdown of carbohydrates and nitrogen metabolism. Molybdenum (Mo) helps in the use of nitrogen, and zinc (Zn) is essential for the transformation of carbohydrates into sugars. It is a part of the enzyme systems which regulate plant growth [16].

Except for carbon, oxygen and nitrogen, all other macro and micronutrients are usually available in soil from where plants acquire. While plants acquire carbon and oxygen from the atmosphere through the process of photosynthesis, nitrogen from the atmosphere is available for plants *via* an indirect process called nitrogen fixation into the soil. Macronutrients in the soil get quickly depleted mainly because plants use them in large amounts, and therefore, the soil needs replenishment of macronutrients for intensive cultivation or in case of multiple cultivations in quick succession. In the case of rice cultivation, it is usual to replenish the soil with fertilizers containing a soluble form of potassium, nitrogen and phosphorus. Here in our experiment, we replenished soil with SSP to provide phosphorus and urea to replenish the soil with nitrogen, but to replenish the soil with potassium banana plant pseudo-stem juice was used in lieu of MOP. The secondary macronutrients such as calcium (Ca) and magnesium (Mg) are detected in comparatively less amount in the juice. There are usually enough of these nutrients in the soil, so fertilization is not always needed. The presence of a considerable amount of K^+ in the banana pseudo-stem juice makes it useful to apply in agriculture as fertilizer in place of MOP.

Photographs in Figs. (**1** to **5**) do show the gradual developments of paddy plants. From the photographs, it is clear that in treatments T2, T3and T4, saplings are developing faster and greener as compared to those in treatment T1. In the initial stage of 60-65 days, the growth of saplings in treatment T4 appeared as good as those of saplings in treatments T2 and T3. However, saplings in T4 started slowing down their growth after about two months, while those in T2 and T4 continued to grow as usual. Growth of saplings in T1 was slow from the very beginning as compared to those in treatments T2, T3 and T4, and finally, harvesting was delayed by about a week.

Fig. (1). Paddy after 15 days of transplantation.

Slow growth of saplings in T1 is obviously due to the absence of adequate quantities of NKP fertilizers in normal agricultural soil. No NKP supplement from external sources was added in treatment T1. Although no potash supplement from an external source was added in treatment T4, saplings continued to grow normally in the initial stage with NP supplements from external sources using potash available in normal agricultural soil. Inadequate supply of potash in the later stage must be responsible for the slowdown of growth of saplings in treatment T4. As adequate NKP supplements from external sources were provided in treatments T2 and T3, saplings in T2 and T3 continued to grow normally. It is noteworthy that the growth of saplings in both T2 and T3 appears identical, thus indicating that as far the growth of saplings is concerned, banana plant juice is no inferior to MOP. Better development in T4 as compared to T1 is due to the effect of either urea or SSP or both. None of the two fertilizers was added to T1.

Fig. (2). Paddy after 65 days of transplantation.

Yields of rice from each treatment have been analyzed with respect to the number of seeds per panicle, weight of seeds and total yield per treatment, and the results are shown in Table **5**. Results show that the average number of seeds per panicle in normal soil is 107.7 (T1, control) against 266.6 (treatment T2) in soil treated with banana plant juice (BPJ) and 222.2 (treatment T3) in soil treated with MOP. This implies that both BPJ and MOP are performing better than the control, the count of seeds per panicle in both cases being more than double as compared to that in control. If treatments T2 and T3 are compared, it is observed that the number of seeds per panicle in treatment with BPJ (treatment T2) is significantly higher as compared to that with MOP (treatment T3). Number of seeds per panicle in the treatment T3 is 222.2 against 195.2 in the treatment T4. In both T3 and T4, equal quantities of urea and SSP were applied, but in T3 MOP was added but no MOP in T4. This implies that urea and SSP perform better in the presence of potassium. The number of seeds in different treatments is graphically shown in Fig. (**6**).

Fig. (3). Paddy after 90 days of transplantation.

Fig. (4). Paddy after 110 days of transplantation.

Fig. (5). Paddy after 120 days of transplantation.

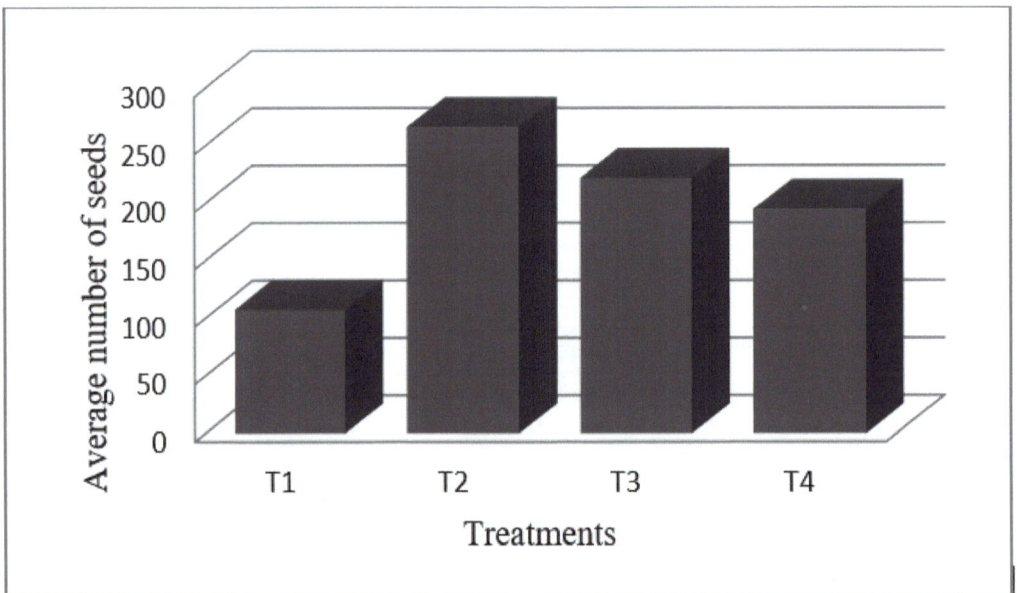

Fig. (6). Average number of rice seeds per panicle in different treatments.

Table 5. Average number of seeds per panicle, average weight per 300 seeds and total yield of rice in different treatments.

Entry	Treatment	Average Number of Seeds/Panicle	Average Weight/300 Seeds (g)	Weight/plot (kg)	Weight/ha (kg)
1	T1	107.7 ± 2.6^a	5.03 ± 0.02^a	0.452	3040.8
2	T2	266.6 ± 5.6^b	5.66 ± 0.02^b	1.236	8315.1
3	T3	222.2 ± 4.6^c	5.50 ± 0.01^c	1.118	7521.2
4	T4	195.2 ± 3.9^d	5.27 ± 0.02^d	0.878	5906.7
	F Test	*	*		

Value for average number of seeds/panicle represents mean ± standard error of ten replicates whereas the value for average weight of seeds /300 seeds represents mean ± standard error of three replicates.
F test: *: $P < 0.05$
a, b, c and d: Means followed by the same letter are not significantly different according to Tukey's method at 5% level.

Average weight of 300 seeds in different treatments can be seen in Table **5**. It is observed that average weight of 300 seeds in T2 is highest (5.66 g) followed by T3 (5.50 g) and T4 (5.27 g). Least weight is observed in T1 (5.03) where none of NPK fertilizers was applied. From these results, it is clear that soil treated with BPJ offers not only a higher number of seeds per panicle or per plant but also healthier seeds as compared to soil treated with MOP. From the comparison of seed weight in T3 (5.50 g) with that in T4 (5.27 g), it appears that urea and SSP are performing better in the presence of MOP. A graphical representation of the average weight of 300 seeds from each treatment is shown in Fig. (**7**).

Total yield of rice in T2 is 1236 g against 1118 g in T3, thus showing an improvement of 10% of yield in treatment of soil with BPJ over that with MOP. Performance of BPJ (T2, 1236 g) over that of normal soil (T1, 452 g) is almost three times. Total yield in both T2 and T3 is significantly higher as compared to that in T4 (878 g). This again implies that urea and SSP can perform better in the presence of potassium. A graphical presentation of the results is shown in Fig. (**8**).

The statistical analysis on the average number of seeds (rice) per panicle (Table **5**) shows that there exists a significant difference ($P< 0.05$) among the different treatments. Further, the critical difference value along with the average number of seeds per panicle under different treatments indicates that treatments T1, T2, T3 and T4 are significantly different from each other. The average value observed indicates that the best result was obtained with treatment T2, where banana pseudo-stem juice was applied in lieu of muriate of potash followed by T3 the treatment with MOP.

Seed weight is a major determinant of the yield of rice. The effect of treatments on the average weight per 300 seeds shows that there also exists a significant difference among the treatments. The critical difference value and the average weight per 300 seeds show that all four treatments are significantly different from each other. The highest average value was observed in the treatment T2 followed by T3.

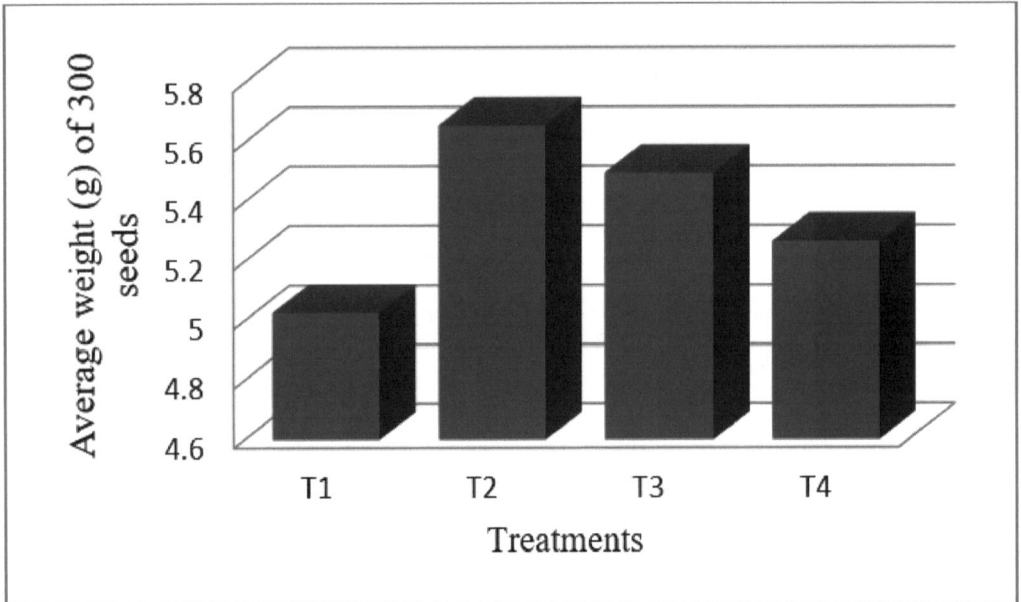

Fig. (7). Average weight of 300 paddy seeds in different treatments.

Treatment T2 (application of banana plant pseudo-stem juice) produced the highest average seed weight and highest average number of seeds per panicle, followed by treatment T3. In T2 and T3, N and P fertilizers were applied from the same chemical sources, but the source of potassium was different. The higher average value with treatment T2 may be due to the extra nutrients present in banana plant pseudo-stem juice. Significantly lower average seed weight and average number of seeds per panicle found with treatment T1 must be due to insufficient nutrition. In the treatment T4, the seed weight was less than that with treatments T2 and T3. This may be due to lack of available potassium. The different average weight of 300 seeds and average number of seeds per panicle were the causes of different yields in the treatments. Therefore, the highest yield was obtained in the treatment T2 followed by T3, T4 and least in treatment T1. It should be noted that Zn and Fe deficiencies in agricultural soil have been reported in India [20]. Therefore, the presence of Fe and Zn in banana pseudo-stem juice (Table **4**) makes it worthy to use in agriculture to increase soil fertility.

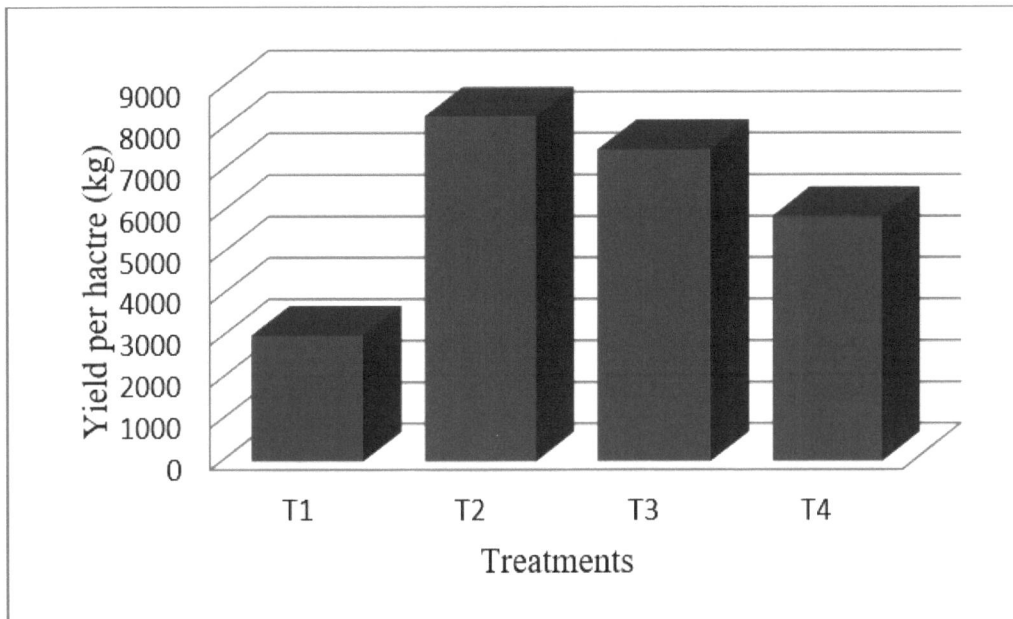

Fig. (8). Yield of rice per hectare in different treatments.

In summary,

- Highest average number of seeds is recorded in the treatment T2 (with BPJ) followed by T3 (with MOP), T4 (no K), T1 (no NPK).
- Highest average weight of seeds is recorded in the treatment T2 followed by T3, T4, T1.
- Highest total yield also obtained in the treatment T2 followed by T3, T4, T1.

4. CONCLUSION

Post-harvest banana plants are a waste. The juice of pseudo-stem is rich in potassium along with a host of micronutrients. Application of banana plant pseudo-stem juice is significantly effective in the production of rice. Application of banana plant pseudo-stem juice in lieu of muriate of potash affords higher yield of rice production. It is observed that both the number of seeds per panicle as well as the average weight of seeds is higher with banana plant pseudo-stem juice as compared to corresponding values with MOP. Thus, banana plant pseudo-stem juice is an ideal replacement or even better replacement for MOP in the cultivation of rice.

REFERENCES

[1]   a) Orovin, T.L.; McFarland, M.L. *Essential Nutrients for Plants*; Texas A & M Agrilife Extension: USA, **2020**. www.agrilifees tension.tamu.edu/library/gardening/essential-nutrients-for-plants/accessed 2020-06-10. b) *Introduction to Plant Science,* **2015**, 88-102. www.ncagr.gov/cyber/kidswrld/plant/nutrient.htm

[2]   a) Manoharan, N. *SOil Test Based Integrated Nutrient Tailoring for Optimum Banana Production and Sustainable Soil Health using Artificial Neural Networks (Ph.D. Thesis)*; Vinayaga Mission's University: Salem (Tamil Nadu, India), **2012**. b) Raghuveer, ; Baghel, S.S.; Puri, G.; Aher, S.B; Jatav, R.C. Targeted yield concept based fertilizer recommendation for garlic (*Allium sativum* L.) in black soil of Madhya Pradesh. *Int. J. Pure App. Biosci,* **2017**, *5*(3), 678-689. [http://dx.doi.org/10.18782/2320-7051.2691]. c) Sharma, G.K.; Mishra, V.N.; Sankar, G.R.M.;Patil, S.K.; Srivastava, L.K.; Thakur, D.S.; Rao, C.S. Soil-test-based optimum fertilizer doses for attaining yield targets of rice under Midland Alfisols of Eastern India. *Commun. Soil Sci. Plant Anal.,* **2015**, *46*, 2177-2190. [http://dx.doi.org/10.1080/00103624.2015.1069319]. d) Dey, P. Targeted yield approach of fertilizer recommendation for sustaining crop yield and maintaining soil health. *JNKVV Res. J.,* **2015**, *49*(3), 338-346.

[3]   a) Henkel, M. *21ˢᵗ Century Homestead: Sustainable Agriculture III: Agricultural Practices,* **2020**, accessed 2020-06-10. b) Erisman, J.W.; Sutton, M.A.; Galloway, J.; Klimont, Z.; Winiwarter, W. How a century of ammonia synthesis changed the world. *Nat. Geosci.,* **2008**, *1*, 636-639. www.nature.com/naturegeoscience [http://dx.doi.org/10.1038/ngeo325]

[4]   a) Joshi, M. *New Vistas of Organic Farming,* 2ⁿᵈ ed; Scientific Publishers: (India): Delhi, **2016**. b) Bhardwaj, K.K.R. *Recycling of Crop, Animal, Human and Industrial Wastes in Agriculture*; Tandon, H.L.S., Ed.; Pub Fertilizer Development & Consultation organization: New Delhi, **1995**, pp. 9-27. c) Mondal, D.; Sinha, S.K. Recycling of organic wastes in agriculture through vermicompost and its significance on environment. *Asian J. Environ. Sci.,* **2012**, *7*(2), 251-254.

[5]   Enwall, K.; Philippot, L.; Hallin, S. Activity and composition of the denitrifying bacterial community respond differently to long-term fertilization. *Appl. Environ. Microbiol.,* **2005**, *71*(12), 8335-8343. [http://dx.doi.org/10.1128/AEM.71.12.8335-8343.2005] [PMID: 16332820]

[6]   a) Lal, R. Soil carbon sequestration impacts on global climate change and food security. *Science,* **2004**, *304*(5677), 1623-1627. [http://dx.doi.org/10.1126/science.1097396] [PMID: 15192216]. b) *Proceedings of the Global Symposium on Oil Organic Carbon,* **2017**, 21-23 March 2017.c) Rees, E. *Change farming to cut CO²* *emissions by 25 per cent,* **2009**, https://theecologist.org/2009/jul/03/change-farming-cut-co2-em issions-25-cent

[7]   a) Pimentel, D.; Hepperly, P.; Hanson, J.; Douds, D.; Seide, R. Environmental, energetic, and economic comparisons of organic and conventional farming systems. *Bioscience,* **2005**, *55*(7), 573-582. [http://dx.doi.org/10.1641/0006-3568(2005)055[0573:EEAECO]2.0.CO;2]. b) Mäder, P.; Fliessbach, A.; Dubois, D.; Gunst, L.; Fried, P.; Niggli, U. Soil fertility and biodiversity in organic farming. *Science,* **2002**, *296*(5573), 1694-1697. [http://dx.doi.org/10.1126/science.1071148] [PMID: 12040197]

[8]   a) Singh, B. *Integrated Plant Nutrient Management*; Farming Outlook, **2018**, pp. 2-7. b) Priyadharshini, J.; Seran, T.H. Paddy husk ash as a source of potassium for growth and yield of cowpea (*Vignaunguiculata* L.). *J. Agric. Sci.,* **2009**, *4*(2), 67-76. c) Htun, K.M.; Mar, S.S.; Thein, S.S.; Toe, K.; Ngwe, K. Effects of different rates of potassium fertilizer on rice productivity with or without rice husk ash in Minbya soil. *J. Agric. Res. (Lahore),* **2017**, *4*(1), 30-38.

[9]   Potassium fertilization in crop production. *In: Soils, Fertility and Nutrients; Agriculture Knowledge Centre,* https://www.saskatchewan.ca/business/agriculture-natural-resources-and-industry/agribus iness-farmers-and-ranchers/crops-and-irrigation/soils-fertility

[10]   Kumar, P.; Sharma, M.K. *Nutrient Deficiencies of Field Crops: Guide to Diagnosis and Management*; CAB International: Oxfordshire, UK, **2013**.
[http://dx.doi.org/10.1079/9781780642789.0000]

[11]   a) *Global Potassium Reserves and Potassium Fertilizer Use,* **2008**, October 6. b) *Mineral Commodity Summaries 2019*; U.S. Geological Survey: Reston, Virginia, **2019**.

[12]   Fite, R.C. *Origin and Occurrence of Commercial Potash Deposits*; Academy of Science, **1951**, pp. 123-125.

[13]   Oliveira, L.; Cordeiro, N.; Evtuguin, D.V.; Torres, I.C.; Silvestre, A.J.D. Chemical composition of different morphological parts from 'Dwarf Cavendish' banana plant and their potential as a non-wood renewable source of natural products. *Ind. Crops Prod.,* **2007**, *26*, 163-172.
[http://dx.doi.org/10.1016/j.indcrop.2007.03.002]

[14]   a) Neog, S.R. *Banana Plant: A Renewable Source of Potassium Chloride and Potassium Carbonate (Ph.D. Thesis),* **2014**. b) Deka, D.C.; Talukdar, N.N. Chemical and spectroscopic investigation of *kolakhar* and its commercial importance. *Indian J. Tradit. Knowl.,* **2007**, *6*(1), 72-78.

[15]   a) Potassium in plants and soil. *In: Essentials of Fertilizers & Irrigation Management,* **2020**, https://www.smart-fertilizer.com/articles/potassium-in-plants. b) Potassium *In: Crop Nutrients,* **2020**, https://www.cropnutrition.com/

[16]   Dubey, K. *Plant Nutrients and it's Role,* **2011**, http://agropedia.iitk.ac.in/content/plant- nutrients-an--its-role

[17]   Rout, G.R.; Sahoo, S. Role of iron in plant growth and Metabolism. *Rev. Agric. Sci.,* **2015**, *3*, 1-24.
[http://dx.doi.org/10.7831/ras.3.1]

[18]   a) White, P.J.; Broadley, M.R. Calcium in plants. *Ann. Bot.,* **2003**, *92*(4), 487-511.
[http://dx.doi.org/10.1093/aob/mcg164] [PMID: 12933363]. b) *The importance of calcium,* www.tetrachemicals.com/Products/Agriculture/The_Importance_of_Calcium

[19]   a) Hauer-Jákli, M.; Tränkner, M. Critical leaf magnesium thresholds and the impact of magnesium on plant growth and photo-oxidative defense: A systematic review and meta-analysis from 70 years of research. *Front. Plant Sci.,* **2019**, *10*, 766.
[http://dx.doi.org/10.3389/fpls.2019.00766] [PMID: 31275333]. b) Guo, W.; Chen, S.; Hussain, N.; Cong, Y.; Liang, Z.; Chen, K. Magnesium stress signaling in plant: just a beginning. *Plant Signal. Behav.,* **2015**, *10*(3), e992287.
[http://dx.doi.org/10.4161/15592324.2014.992287] [PMID: 25806908]

[20]   a) Palanog, A.D.; Calayugan, M.I.C.; Descalsota-Empleo, G.I.; Amparado, A.; Inabangan-Asilo, M.A.; Arocena, E.C.; Sta Cruz, P.C.; Borromeo, T.H.; Lalusin, A.; Hernandez, J.E.; Acuin, C.; Reinke, R.; Swamy, B.P.M. Zinc and iron nutrition status in the Philippines population and local soils. *Front. Nutr.,* **2019**, *6*, 81.
[http://dx.doi.org/10.3389/fnut.2019.00081] [PMID: 31231657]. b) Reddy, S.R. *Agronomy of field crops,* 3rd ed; Kalyani Publishers: Ludhiana, India, **2009**. c) Singh, M.V. Micronutrient deficiencies in crops and soils in India. In: *Micronutrient Deficiencies in Global Crop Production*; Alloway, B.J., Ed.; Springer: UK, **2008**.
[http://dx.doi.org/10.1007/978-1-4020-6860-7_4]

CHAPTER 4

Banana Plant Pseudo-stem Juice: a Better Substitute for MOP in Wheat Cultivation

Abstract: Banana plant pseudo-stem juice is a rich source of potash and used in the cultivation of wheat. Use of the juice in lieu of muriate of potash affords a higher number of seeds per plant as well as a higher per capita weight of seeds. The juice treated soil affords 60% more yield over that in soil treated with muriate of potash; 89% more as compared to the yield from normal soil not treated with muriate of potash. Materials and methods have been discussed. Colour pictures of experimental plots are reported.

Keywords: Banana plant juice, Banana plant pseudo-stem, Organic potash, Substitute for MOP, Wheat cultivation.

1. INTRODUCTION

Wheat is the third most important global staple food crop grown worldwide in different climatic conditions, but its optimal growing conditions are in temperate environments [1]. There are many countries where wheat is the main cash crop. It is the dominant crop in approximately 90% arid and semiarid and rain-fed areas of Iran [2]. South Australia produces protein-rich wheat, which is used to produce high-quality bread [3], and Australia ranks fourth in the world in wheat exports.

Wheat is a globally traded food commodity, and it is one of the major sources of calorie in human diets. To ensure food security, balancing the production and consumption of wheat is essential. Approximately 70% of global wheat production is used for food [4], and the demand is rising due to population growth, economic cxdevelopment, increasing wealth, urbanization and changing lifestyles [5]. It is possible to increase wheat production substantially by intensifying the use of fertilizer and irrigation in the farming lands currently being used for wheat cultivation [6].

2. EXPERIMENTAL DETAILS

2.1. Materials

The wheat variety, *Triticum aestivum* (sunalika, the irrigated late variety), was used in the experimental cultivation. The rest of this section is the same as in Section 2.1 of Chapter 3.

2.2. Details of Methodology

For wheat cultivation, four plots of lands, each of size 4ft × 4ft, were used. The soil was tested in the laboratory of the Department of Agriculture, Lakhimpur district, Assam (*cf.* Table **1** in Chapter 3), and based on the test report, the plots were prepared for cultivation. Recommended quantities of fertilizers required for the cultivation of wheat are shown in Table **1**. All four plots were given different treatments, as shown in Table **2**. In treatment T2 (Plot 2), banana plant juice as a source of potassium was applied to the soil as a basal application in lieu of commercial potash and in the treatment T3 (Plot No. 3) muriate of potash (MOP) was used. No potash was applied in the treatment T4 (Plot No. 4). Calculated quantities of urea and single super phosphate (SSP) were applied in all the treatments (Plots 2, 3, and 4) except the treatment T1 (Plot 1). Treatment T1 was control.

Table 1. Recommended fertilizers (in kg/bigha, 7.47 bigha = 1 hectare) for wheat cultivation [variety, *Triticum aestivum* (sunalika, the irrigated late variety)].

Farm Yard Manure (FYM)	= 1500 kg
Urea	= 28.0 kg
Single Super Phosphate (SSP)	= 21.3 kg
Muriate of Potash (MOP)	= 7.2 kg
Lime	= not required

Table 2. Fertilizers applied to different plots for wheat cultivation (g/plot).

Plot	Treatments	Urea (g)	SSP (g)	MOP (g)	Banana Plant Juice (Litres)
1	T1	Nil	Nil	Nil	Nil
2	T2	31	24	Nil	1.800
3	T3	31	24	8	Nil
4	T4	31	24	Nil	Nil

Procedure: Four plots, each of size 4ft x 4ft were used for experimental purposes. The plots were parts of a piece of normal agricultural land. The plots, in dry condition, were ploughed several times to bring the soil into fine tilth. The plots were labeled as T1, T2, T3 and T4 for the convenience of experimental recording. During ploughing, 1.7 kg of FYM was spread uniformly to each of the plots and got mixed well in the soil. In the final stage of land preparation, the whole amount of superphosphate, MOP and banana plant juice as shown in Table **2** were applied to the soil of the corresponding plots and mixed well. One-third of urea was applied as basal in the final stage of land preparation, one-third after 15 days sowing seeds, and the remaining one-third after 30 days of sowing seed. In each plot, 18.0 g of wheat seeds was broadcasted in four rows at an equal distance apart. Sowing was done in the last week of the month of December, which is the normal time of wheat cultivation in this part of India. From the date of sowing until harvesting, in total, the plots were irrigated six times. Irrigation schedule, as per advice from experts in Agriculture, was as follows:

First Irrigation: The first irrigation was given after 20 days of sowing, *i.e.*, at the crown root initiation stage.

Second Irrigation: It was given at the tillering stage, 35 days after sowing.

Third Irrigation: At the late jointing stage, 55 days after sowing.

Fourth Irrigation: Given at flowering stage 65 days after sowing.

Fifth Irrigation: Milk formation stage 75 days after sowing.

Sixth Irrigation: Given at dough stage, 85 days after sowing.

2.3. Collection of Seeds

In treatments T2, T3 and T4, it took 102 days for seeds to mature against 109 days in T1. Seeds were collected in multiple ways so that these could be compared as average weight per seed in different treatments, number of seeds per panicle or plant in different treatments and total yield from different treatments. From each treatment, 10 panicles were randomly collected, and the numbers of seeds in each panicle were counted. Further, 300 seeds from each treatment were randomly collected and weighed. And finally, the total yield in each treatment was recorded.

3. STATISTICAL ANALYSIS

All the collected data were statistically analysed by analysis of variance (ANOVA) and Tukey's method at 5% level was applied for multiple comparisons between two average values in the treatments. 'F Test' is one of the best methods for differences among more than two average values of different treatments. With the help of 'F Test', attempt was made to observe differences between the average values of different treatments shown in Table **3**.

4. RESULTS AND DISCUSSION

Macro and micro nutrients present in banana plant pseudo-stem juice are discussed in Section 3.3, Chapter 3 (*cf.* Chapter 2 for methods of estimation).

Plants acquire their essentials nutrients from soil except for carbon and oxygen, which are available in the air. All macro and micro nutrients, except nitrogen, are absorbed from the soil. Nitrogen is available to plants from the atmosphere *via* an indirect process called nitrogen fixation into the soil. Macronutrients such as potassium, nitrogen and phosphorus in the soil get quickly depleted mainly because plants use them in large amounts, and therefore, periodical replenishment of soil with these macronutrients are essential for intensive cultivation or in case of multiple cultivations in quick successions. It is usual to replenish the soil with potassium, nitrogen and phosphorus in their water-soluble forms. In our experiment, we used SSP to replenish the soil with phosphorus and urea to replenish the soil with nitrogen, but to replenish the soil with potassium banana plant pseudo-stem juice is used in lieu of MOP for the purpose of experiment. The secondary macronutrients such as calcium and magnesium are usually not required in fertilizer as enough of these are present in the soil. Because of the presence of the considerable amount of K^+ in the banana plant pseudo-stem juice, we have considered it as a suitable substitute for potash in crop cultivation.

Experimental wheat cultivation (variety *Triticum aestivum*) is shown in Figs. (**1** to **7**). It is evident from the photographs that seedlings in treatments T2 and T3 are developing faster as compared to those in treatments T1 and T4. Seedling is also healthier and greener in T2 and T3 treatments. In treatment T2, banana plant juice (BPJ) has been used, while MOP has been used in treatment T3. Looking at the seedlings, one should be convinced that BPJ is as good as MOP, if not better. Slower developments of seedlings in T1 and T4 are due to the inadequate presence of potassium. Neither banana plant juice nor muriate of potash has been used in these two treatments. Better development of seedlings in T4 as compared to T1 is due to the effect of urea and SSP. None of these two fertilizers was used in T1. Harvesting of treatment T1 was delayed by a week.

Yields of wheat from all four treatments are shown in Table **3**. Yields may be compared with respect to the number of seeds per panicle or plant, average seed weight and total yield. In normal soil (treatment T1, control), the average number of seeds per panicle is 16.9 against 48.7 in banana plant juice (BPJ) treated soil (treatment T2) and 37.4 in MOP treated soil (treatment T3). This implies that potassium helps to increase the number of seeds, and organic potassium from banana plant juice works better than muriate of potash. The number of seeds per panicle (Figs. **1** to **7**) in BPJ treated soil is nearly 3 times that in control (T1). As compared to control, the number of seeds per panicle in MOP treated soil is double. If the number of seeds per panicle in treatments T2 and T3 are compared, it is observed that the number of seeds per panicle in BPJ treated soil (treatment T2) is significantly higher as compared to that in MOP treated soil (T3). Thus, the performance of banana plant juice as fertilizer is better than MOP. The average number of seeds per panicle in treatment with MOP (treatment T3) is 37.4 against 32.2 in treatment without MOP (treatment T4). In both T3 and T4, equal quantities of urea and SSP were used; MOP was used in T3 but not in T4. This implies that urea and SSP perform better in the presence of potassium. A graphical presentation of the average number of seeds per panicle in different treatments is shown in Fig. (**8**).

Table 3. Average number of seeds per panicle, average weight per 300 seeds and total yield of wheat in different treatments.

Entry	Treatment	Average Number of Seeds per Panicle	Average Weight per 300 Seeds (g)	Yield per Plot (kg)	Yield per Hectare (kg)
1	T1	16.9 ± 0.67^a	10.17 ± 0.04^a	0.252	1695.3
2	T2	48.7 ± 1.87^b	12.33 ± 0.04^b	0.620	4171.0
3	T3	37.4 ± 1.67^c	11.42 ± 0.06^c	0.388	2610.2
4	T4	32.2 ± 1.49^d	11.17 ± 0.04^d	0.328	2206.6
	F Test	*	*		

Value for average number of seeds per panicle represents mean ± standard error of ten replicates whereas the value for average weight per 300 seeds represents mean ± standard error of three replicates. F test: *: $P < 0.05$ a, b, c and d: Means followed by the same letter are not significantly different according to Tukey's method at 5% level.

Average total weight of randomly selected 300 seeds from different treatments are recorded in Table **3**. The highest weight for 300 seeds is observed in treatment T2 (12.33 g) followed by treatment T3 (11.42 g) and treatment T4 (11.17 g). The least weight is observed in treatment T1 (10.17). The results imply that soil treated with banana plant juice offers not only a higher number of seeds per plant but also a higher per capita weight of seeds as compared to soil treated with muriate of potash. If seed weight in treatment T3 (11.42 g) is compared with that in treatment T4 (11.17), it appears that urea and superphosphate as fertilizer do perform better in the presence of potash. Least weight in treatment T1 is observed because none of the major nutrients *viz*. nitrogen, phosphorus and potassium fertilizers were used in this treatment. The total average weights of 300 seeds in different treatments are graphically shown in Fig. (**9**).

Fig. (1). After 10 days of sowing.

Fig. (2). After 25 days of sowing.

Fig. (3). After 40 days of sowing.

Fig. (4). After 64 days of sowing.

Fig. (5). After 72 days of sowing.

Fig. (6). After 86 days of sowing.

Fig. (7). After 102 days of sowing.

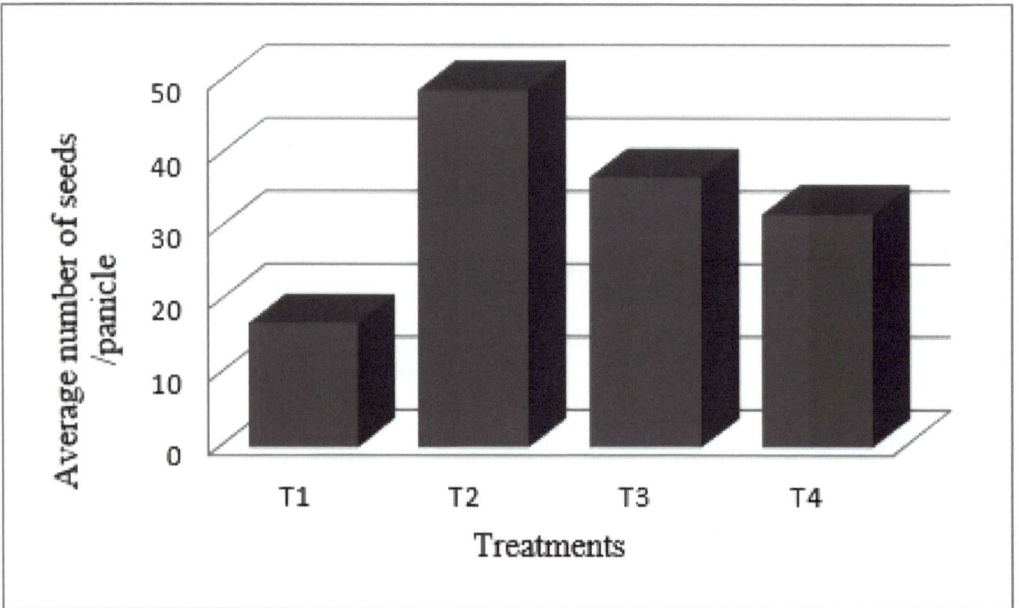

Fig. (8). Average number of wheat seeds per panicle in different treatments.

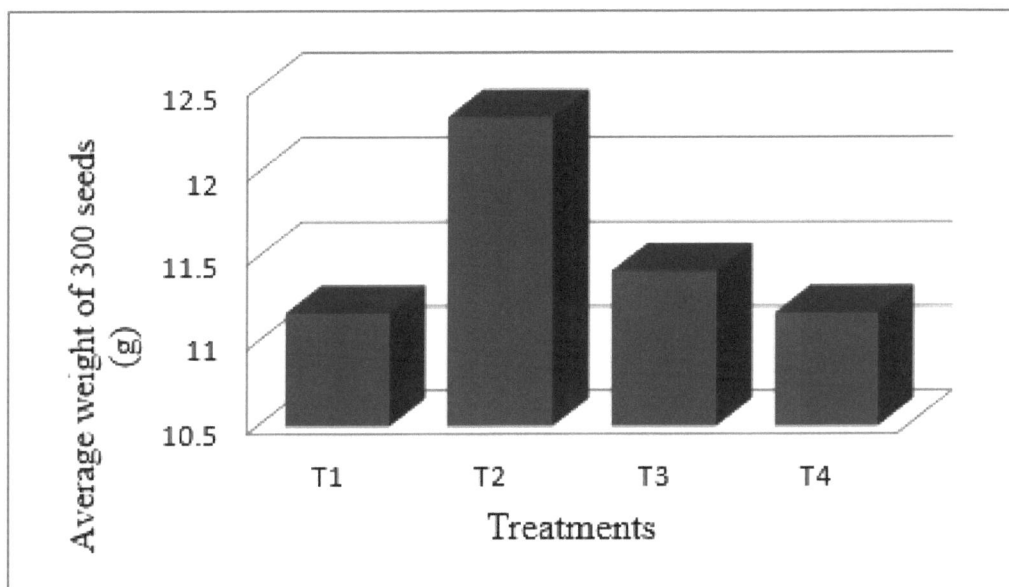

Fig. (9). Average weight of 300 wheat seeds in different treatments.

Highest yield of wheat (620 g) is observed in treatment T2 (soil treated with BPJ) and least (252 g) in treatment T1 (control, no fertilizer used). Yield in soil treated with BPJ shows an improvement of nearly 60% over that (388 g) in soil treated with MOP. Corresponding improvement in rice cultivation was only 10% (*cf.* Chapter 3). Performance of BPJ in wheat cultivation is nearly six times better as compared to that in rice cultivation. It is amazing to observe that performance of BPJ over MOP is much better in wheat cultivation as compared to that in rice cultivation (*cf.* Chapter 3, Section 3). Total yield in treatment T3 (388 g) is significantly higher as compared to that in treatment T4 (328 g). This is expected because all the three major fertilizers, namely potash, urea and SSP were used in the treatment T3, and they together performed better. In treatment T4, only two, namely urea and SSP were used. Results are graphically shown in Fig. (**10**).

Treatment T2 (banana plant pseudo-stem juice used) produced seeds with the highest average weight and panicles with highest average number of seeds followed by treatment T3. In both the treatments, T2 and T3, same nitrogen and phosphorus fertilizers were used, urea and SSP, respectively. But the sources of potassium were different, banana plant juice in T2 and muriate of potash in T3. Better results with the treatment T2 may be due to better availability of potash from banana plant pseudo-stem juice and the extra nutrients present in it. Significantly poor results with treatment T1 (control) must be due to an insufficient supply of nutrients. Poor results with the treatment T4 as compared to treatments T2 and T3 must be due to poor availability of potassium. This indicates the importance of potassium in the enhancement of crop yields.

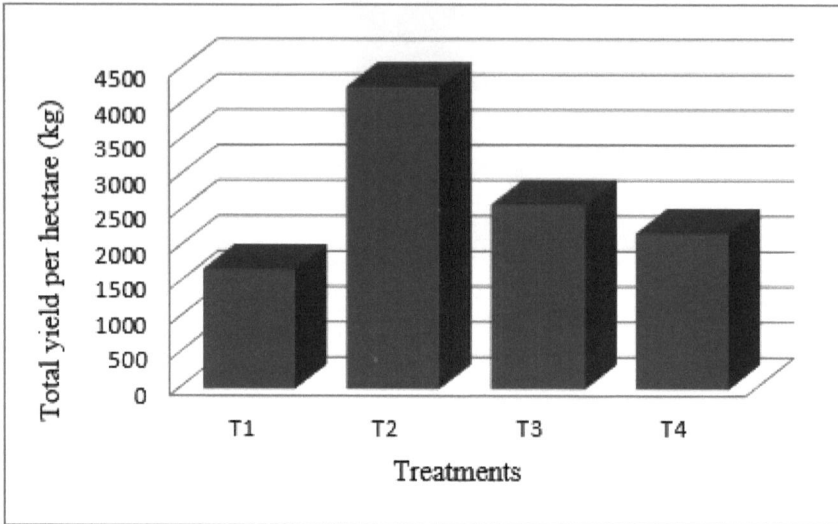

Fig. (10). Yield of wheat per hectare in different treatments.

Statistical analysis of results from different treatments shown in Table **3** is done using 'F Test' of variance (ANOVA) and searched for the pairwise differences. Analysis shows a significant difference (P<0.05) among the different treatments. Treatments T1, T2, T3 and T4 are significantly different from each other. The average value observed indicates that the best result was obtained with treatment T2 where banana plant pseudo-stem juice was applied in lieu of muriate of potash followed by T3 the treatment with MOP.

5. CONCLUSION

The juice of post-harvest banana plant pseudo-stem is highly rich in potassium, also it contains some other micro nutrients. Its use in wheat cultivation is significantly effective. Use of the juice in lieu of muriate of potash has afforded 60% higher yield of wheat. A number of seeds per plant as well as the per capita weight of seeds, are also higher when banana plant pseudo-stem juice is applied in lieu of potash. The possible reason for better performance by banana plant pseudo-stem juice is the better availability of organic potassium to plants. Banana plant pseudo-stem juice is an ideal substitute or even better replacement for muriate of potash in the cultivation of wheat. When compared with rice cultivation, the performance of banana plant juice in wheat cultivation is much better.

REFERENCES

[1]   Balkovič, J.; van der Velde, M.; Skalský, R.; Xiong, W.; Folberth, C.; Khabarov, N.; Smirnov, A.; Mueller, N.D.; Obersteiner, M. Global wheat production potentials and management flexibility under the representative concentration pathways. *Global Planet. Change,* **2014**, *122*, 107-121. [http://dx.doi.org/10.1016/j.gloplacha.2014.08.010]

[2]   Bannayan, M.; Rezaei, E.E. Future production of rain fed wheat in Iran (Khorasan province): climate change scenario analysis. *Mitig. Adapt. Strategies Glob. Change,* **2014**, *19*, 211-227. [http://dx.doi.org/10.1007/s11027-012-9435-x]

[3]   a) Luoa, Q.; Williams, M.A.J.; Bellotti, W.; Bryan, B. Quantitative and visual assessments of climate change impacts on South Australian wheat production. *Agric. Syst.,* **2003**, *77*, 173-186. [http://dx.doi.org/10.1016/S0308-521X(02)00109-9]. b) Reyenga, P.J.; Howden, S.M.; Meinke, H.; Hall, W.B. Global change impacts on wheat production along an environmental gradient in south Australia. *Environ. Int.,* **2001**, *27*(2-3), 195-200. [http://dx.doi.org/10.1016/S0160-4120(01)00082-4] [PMID: 11697669]

[4]   Röder, M.; Thornley, P.; Campbell, G.; Bows-Larkin, A. Emissions associated with meeting the future global wheat demand: A case study of UK production under climate change constraints. *Environ. Sci. Policy,* **2014**, *39*, 13-24. [http://dx.doi.org/10.1016/j.envsci.2014.02.002]

[5]   a) *FAO Report on 'ExPert Meeting on International Investment in the Agricultural Sector of Developing Countries,* **2011**, 22-23 November. b) Satterthwaite, D.; McGranahan, G.; Tacoli, C. Urbanization and its implications for food and farming. *Philos. Trans. R. Soc. Lond. B Biol. Sci.,* **2010**, *365*(1554), 2809-2820. [http://dx.doi.org/10.1098/rstb.2010.0136] [PMID: 20713386]. c) *FAO Report on 'The future of food and agriculture – Trends and challenges',* **2017**. d) Alexandratos, N.; Bruinsma, J. World agriculture towards 2030/2050: The 2012 revision. *Global Perspective Studies Team: FAO Agricultural Development,* **2012**,

[6]   Mueller, N.D.; Gerber, J.S.; Johnston, M.; Ray, D.K.; Ramankutty, N.; Foley, J.A.; Foley, J.A. Closing yield gaps through nutrient and water management. *Nature,* **2012**, *490*(7419), 254-257. [http://dx.doi.org/10.1038/nature11420] [PMID: 22932270]

CHAPTER 5

Organic Potash in Banana Plant Pseudo-stem Juice: Application in Mustard Cultivation

Abstract: Banana plant pseudo-stem juice is highly rich in potash. Its effectiveness in the cultivation of mustard has been compared with that of muriate of potash. Juice treated soil affords 23% more yield as compared to that in soil treated with potash; 65% more than that in soil not treated with potash. Number of pods per plant as well as average seed weight per 100 pods are also higher if commercial potash is replaced by organic potash available from banana plant pseudo-stem juice. Thus, the banana plant pseudo-stem juice provides a better alternative to commercial potash fertilizer. Colour pictures of cultivation have been reported.

Keywords: Alternative of commercial potash, Banana plant juice, Organic fertilizer, Potash for mustard cultivation, Substitute for MOP, Substitute for potash, Use of banana plant pseudo-stem.

1. INTRODUCTION

Mustard oilseed is the third, after palm and soybean oil, among the major sources of vegetable oil as per data revealed by the National Commodity & Derivatives Exchange Limited (NCDEX) [1]. Among the cultivated oilseeds, its share is approximately 14% in the world's market. Traditionally it is the major cooking oil in several countries across the globe. The oil has a pungent taste and an irritating aroma. It is mechanically extracted by grinding and pressing the seeds. To separate the light pungent fraction, other methods are available [2]. Apart from the use in cooking, other applications of mustard oil include uses in aromatherapy, pharmaceuticals, soap preparation, hair and skin care, *etc*. The light pungent fraction is used in therapeutics. It can stimulate the sweat glands and thus provides benefits to the skin. Mustard oil is also considered an appetizer, anti-bacterial and anti-fungal. Due to its huge consumption in food, the market for mustard oil is vast in the Asia-Pacific region, which includes India, Thailand, and China. The global market for mustard oil is expected to grow significantly. Its use

Dibakar Chandra Deka & Satya Ranjan Neog

has increased significantly in food as well as in other industries, including beverage industry, pharmaceutical industry, personal care industry and cosmetic industry. There are three varieties of mustard oilseeds and these are *Brassica nigra* L. (black mustard), *Brassica juncea* L. (brown mustard), and *Brassica hirta* L. (white mustard). Among the Asian countries, India tops the list in the use and production of mustard oil. Because of the massive production and consumption of *Brassica juncea* variety in India, it is often referred to as Indian mustard [3]. The mustard crop requires temperate climate and moist soil for growth.

2. EXPERIMENTAL DETAILS

2.1. Materials

The brown variety of mustard, *Brassica juncea* (the local variety, M-27), was used. For other details, readers may refer to Section 2.1 of Chapter 3.

2.2. Details of Methodology

The land used in the experiment was normal agricultural land. Four plots of lands, each of size 4ft × 4ft, were used. The soil was tested (*cf.* Table 1, Chapter 3), and based on the test report, quantities of fertilizers required for mustard cultivation were recommended by the Department of Agriculture, Lakhimpur District, Government of Assam (shown in Table **1**). The plots were given different treatments, as shown in Table **2**. In treatment T2, banana plant juice in lieu of muriate of potash was used (plot No. 2). In treatment T3 (plot No. 3), commercial fertilizer MOP was applied. No potash was applied to plot No. 4 (treatment T4). Urea to provide nitrogen and SSP to provide phosphorus were applied equally to plots 2, 3, and 4 (treatments T2, T3 and T4) except plot No. 1 (treatment T1). Plot No.1 (treatment T1) was used as the control, and no fertilizer was applied to this plot.

Procedure: For the experimental cultivation of mustard using banana plant juice, a small plot of normal agricultural land was selected. The selected piece of land, in dry condition, was ploughed several times to bring the soil into fine tilth. It was then divided into four plots, each of size 4 ft x 4 ft and labeled as T1, T2, T3 and T4 for four different treatments. In the final stage of plot preparation, 1.7 kg of FYM was uniformly spread and mixed well with the soil in each of the plots. The whole amount of phosphate (SSP), potash (MOP) and banana plant pseudo-stem juice (BPJ), as shown in Table **2**, was applied to the soil of the corresponding plots in the final stage of land preparation. One-third of urea was applied as basal in the final stage of land preparation. The remaining two-third was applied in two installments, one-third after 15 days of sowing seeds and the last one third after 30 days. In each plot, an equal quantity of seeds (0.80 g) was broadcasted. Sowing was done in the last week of the month of December.

Table 1. Recommended fertilizers for mustard crop* (in kg/bigha, 7.47 bigha = 1 hectare), *Brassica juncea* (the local variety, M-27).

Farm Yard Manure (FYM)	= 1500 kg
Urea	= 28.0 kg
Single Super Phosphate (SSP)	= 30.0 kg
Muriate of Potash (MOP)	= 9.5 kg
Lime	= not required

*Recommended by the Govt. Department of Agriculture, Lakhimpur District, Assam.

Table 2. Fertilizers applied in experimental plots (g/plot).

Plot No.	Treatments Code	Urea (g)	SSP (g)	MOP (g)	Banana Pseudo-Stem Juice (Litres)
1	T1	Nil	Nil	Nil	Nil
2	T2	31	33.3	Nil	2.4
3	T3	31	33.3	10.5	Nil
4	T4	31	33.3	Nil	Nil

Irrigation of the plots was carried out as per the advice of agricultural experts. The first irrigation was done after the germination of seeds in such a way that the soil was just wet. Thereafter, irrigation was done at regular intervals of one week, with special attention to make sure to irrigate just before the flowering started. Plants in treatments T2, T3 and T4 matured after 80 days, and were collected, dried and threshed. In treatment T1, plants took a little longer to mature. These were collected after 85 days, dried and threshed.

2.3. Procedure for Seed Collection

The seed from different plots was collected in such a way that results could be compared in terms of the number of pods per plant, yield per 100 pods and total yield per plot. When the seeds matured, 10 plants from each plot were randomly collected and the average number of pods per plant was counted. Again, 100 pods from each treatment were randomly collected, dried separately under the sun, peeled and seeds were weighed. To assess the total production, all the plants from each treatment were harvested, dried and then threshed. The grains were cleaned and weighed to record the total yield per plot.

3. RESULTS AND DISCUSSION

The methods of estimation for the macro and micronutrients present in banana plant pseudo-stem juice are discussed in Chapter 2 and results are discussed in Section 3, Chapter 3.

Except for carbon and oxygen, all other macro and micronutrients are absorbed by the plants from the soil. Plants use macronutrients in large amounts, and these get quickly depleted. For intensive cultivation or in the case of multiple cultivations in quick succession, fertilizers are required to replenish the soil with macronutrients. In the case of mustard, widely used fertilizers are urea to provide nitrogen, single super phosphate (SSP) to provide phosphorus and muriate of potash (MOP) to provide potassium. Here in our experiment, we have discovered that banana plant pseudo-stem juice is a better replacement for MOP to enrich the soil with potassium. The presence of a considerable amount of K^+ in the banana pseudo-stem juice is noteworthy.

Fig. (1). After 15 days of sowing.

Photographs in Figs. (**1** to **6**) are those of mustard cultivation (variety *Brassica juncea* L.). Treatment T1 is the control in which no fertilizer was used. In treatments T2 and T3, all the three fertilizers, namely nitrogen, phosphorus and potassium, were used. Urea for nitrogen and SSP for phosphorus were used in both the treatments, while MOP was used for potassium in T3 and banana plant pseudo-stem juice was used in lieu of MOP in T2. In T4, urea and SSP were used but no potassium fertilizer was used. Better growth of plants with treatments T2, T3 and T4 as compared to that with T1 is the reflection of the combined effects of fertilizers. Better growth of seedlings with treatments T2 and T3 as compared to that in T4 is the reflection of the effect of potassium on plant health. The poor development of plants with treatment T1 indicates that the normal soil is highly deficient in major plant nutrients such as nitrogen, phosphorus and potassium.

Fig. (2). After 22 days of sowing.

In Tables **3 - 5**, the effects of different treatments on the yield of mustard are compared. Table **3** shows the difference in the number of pods per plant with different treatments, while the effect of different treatments on the weight of mustard seeds from 100 pods is shown in Table **4** and on total yields in Table **5**.

Application of banana plant juice on farming soil brings substantial benefit to mustard cultivation. In normal soil, the number of pods per plant ranges from 8 to 13, while that in banana plant juice applied soil, it ranges from 28 to 34 (Fig. **3**). There is a 3-fold increase in the number of pods per plant in banana plant juice applied soil (treatment T2) as compared to that in normal soil (treatment T1). In MOP applied soil (treatment T3), the number of pods per plant ranges from 26 to 30, and is lower as compared to banana plant juice applied soil. Thus, banana plant juice is a better replacement for MOP to provide potassium to the soil. The number of pods per plant in treatment T4 is 2.8-fold higher as compared to that in treatment T1 (control). This clearly reflects the effect of urea and SSP on the cultivation of mustard. The results are graphically shown in Fig. (**7**).

Fig. (3). After 30 days of sowing.

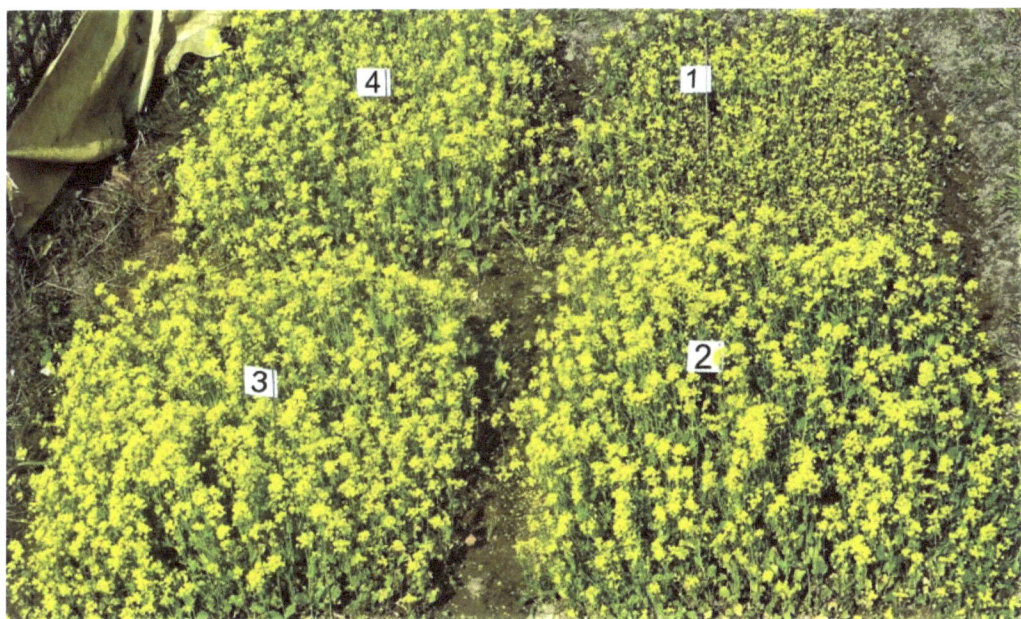

Fig. (4). After 40 days of sowing.

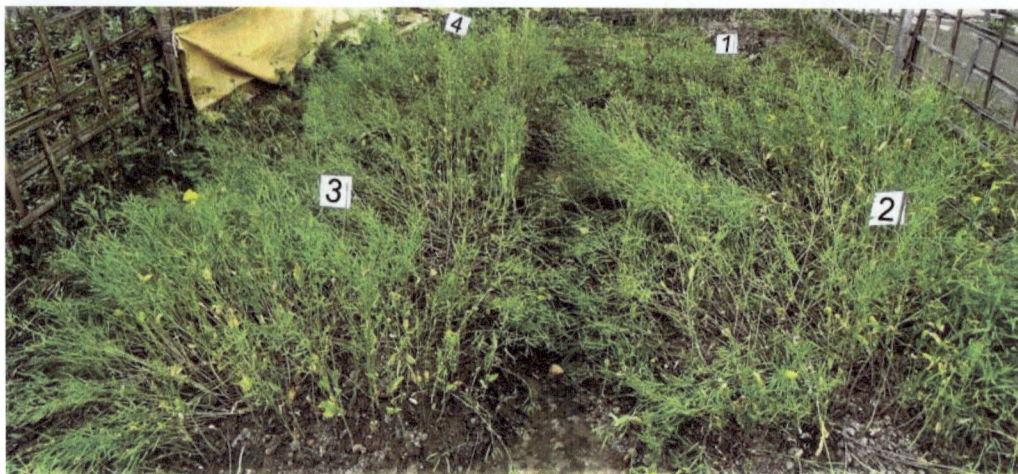

Fig. (5). After 60 days of sowing.

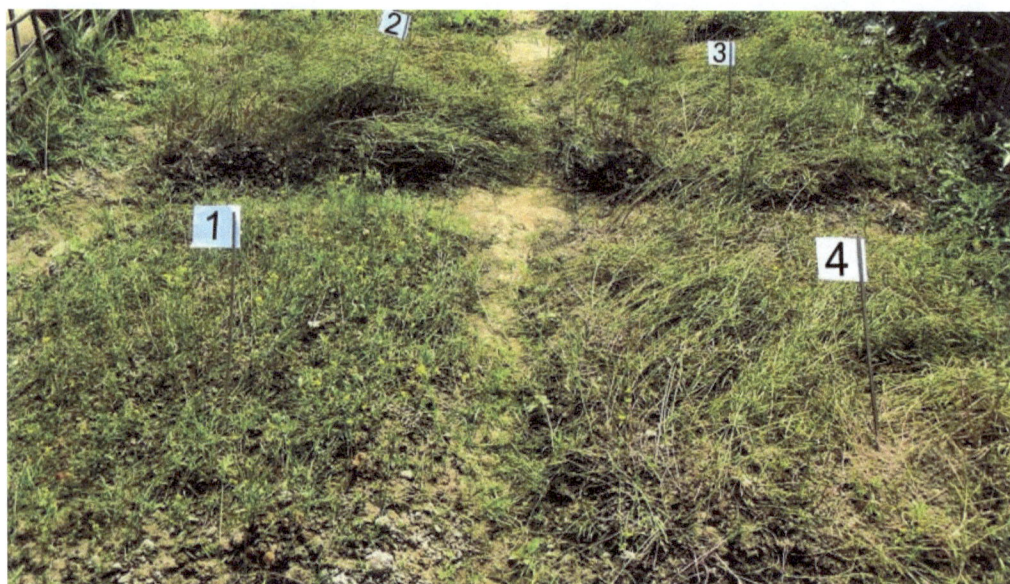

Fig. (6). After 70 days of sowing.

Table 3. Number of pods per plant.

Entry	Total Number of Pods Per Plant			
	T1	T2	T3	T4
1	10	30	30	21
2	8	28	29	24
3	12	34	29	21

(Table 3) cont.....

Entry	Total Number of Pods Per Plant			
	T1	T2	T3	T4
4	12	30	30	22
5	11	28	27	22
6	13	31	26	23
7	12	31	26	24
8	11	30	27	25
9	13	29	28	22
10	13	31	26	18

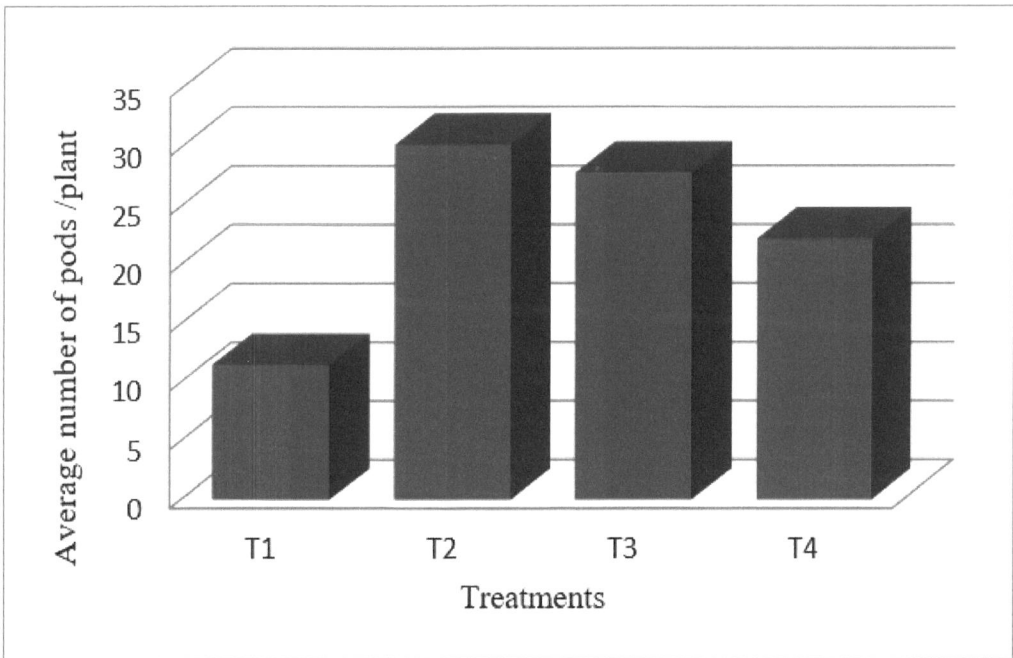

Fig. (7). Average number of pods per plant (mustard) in different treatments.

The total weight of mustard seeds from 100 pods in juice applied soil (T2) is 873 mg against 553 mg in normal soil (T1, Control) and 764 mg in MOP applied soil (T3, Table **4**). Thus, there is a more than 1.5-fold increase of productivity in banana plant juice applied soil as compared to normal soil and more than 1.1-fold increase of productivity as compared to MOP applied soil. Thus, banana plant juice performs better than MOP. The performance of MOP applied soil over normal soil is approximately 1.4-fold. Lower weight (717 mg) of seeds in treatment T4 (with N & P but without K) as compared to that (764 mg) in T3 (with N, P & K) indicates that performance of urea and SSP without potassium is lower by about 6%. Results are shown in Table **4** and graphically presented in Fig (**8**).

Table 4. **Weight of mustard seeds from 100 pods.**

Entry	Weight of Mustard Seeds Per 100 Pods (mg)			
	T1	T2	T3	T4
1	506	843	761	702
2	579	867	792	763
3	573	909	739	686
4 (average of above three)	553	873	764	717

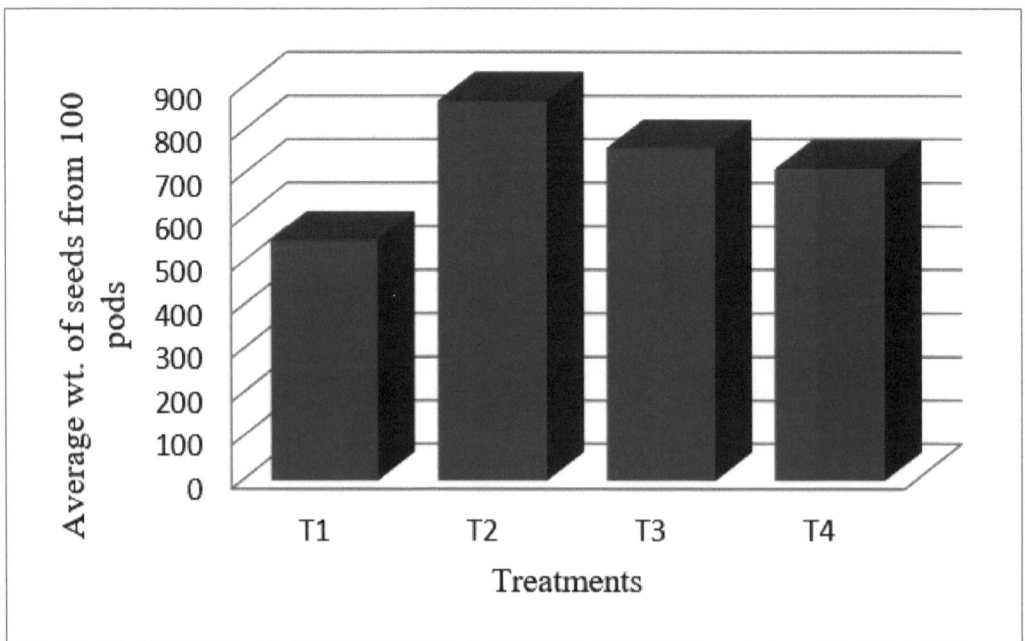

Fig. (8). Average weight of mustard seeds from 100 pods in different treatments.

The total yield of mustard per plot under different treatments is shown in Table **5**. The total yield is 116 g from the plot under treatment T2 against 94 g under treatment T3 and 28 g under treatment T1, thus showing an improvement of more than 23% yield in treatment of soil with banana plant juice over that with MOP, and more than 300% improvement over control (normal soil without fertilizer). In treatment T4, the total yield is 70 g which is lower by about 25% as compared to that under treatment T3 (94 g). Thus potassium can substantially boost the performance of nitrogen and phosphorus fertilizers. Comparison of total yield in T2 with that in T4 indicates an improvement of yield by more than 65%, thus indicating an indispensable role of potassium in the productivity enhancement of soil. The report on yields (Table **5**) shows that the highest yield (780.4 kg/ha) is recorded in the treatment T2 against 632.4 kg/ha in treatment T3, thus an increase of 23% yield with the replacement of MOP by banana plant pseudo-stem juice is achieved. Results are graphically shown in Fig (**9**).

Table 5. Average number of pods per plant, average weight of seeds from 100 pods and total yield of mustard seeds in different treatments.

Entry	Treatment	Average Number of Pods /Plant	Average Weight of Seeds/100 Pods (mg)	Weight/Plot (kg)	Weight/ha (kg)
1	T1	11.5 ± 0.47^a	552.67 ± 19.00^c	0.028	188.4
2	T2	30.2 ± 0.52^b	873.19 ± 15.75^b	0.116	780.4
3	T3	27.8 ± 0.49^c	764.00 ± 12.55^a	0.094	632.4
4	T4	22.2 ± 0.57^d	717.04 ± 19.16^a	0.070	470.9
F Test		*	*		

Average number of pods per plant represents the mean ± standard error of ten replicates. Average weight of seeds per 100 pods represents the mean ± standard error of three replicates. F test: *: $P < 0.05$; a, b, c and d: means followed by the same letter are not significantly different, according to Tukey's method at 5% level.

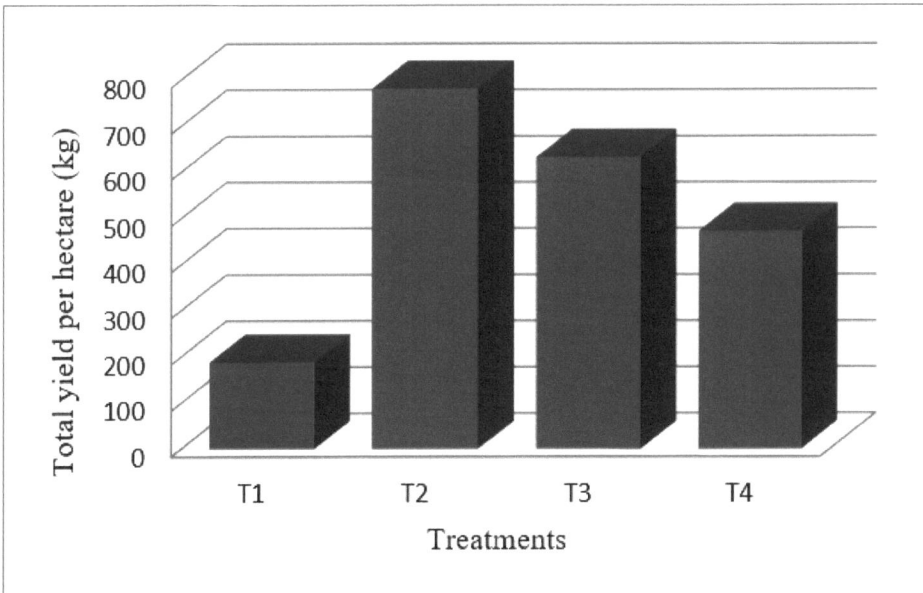

Fig. (9). Yield of mustard per hectare in different treatments.

4. STATISTICAL ANALYSIS

The data collected were statistically analysed by using analysis of variance (ANOVA) and mean separation was done using Tukey's method at 5% level. The average number of pods per plant ranged from 11.5 in treatment T1 to 30.2 in treatment T2, and there is a significant difference in the number of pods among the treatments (Table 5). The differences in the average number of pods per plant under different treatments indicate that the treatments T1, T2, T3 and T4 are significantly different from each other. The highest number of pods is recorded in treatment T2 where banana plant juice was applied in place of MOP, followed by treatment T3, where chemical potash MOP was applied.

Seed weight per pod is a major determinant of the total yield of mustard. The effect of treatments on mustard seed weight per 100 pods (Table 5) indicates that there is a significant difference ($P<0.05$) among the treatments. Treatment with banana plant juice (treatment T2) afforded the highest seed weight followed by treatment T3 where MOP was applied. Significantly lower seed weight per 100 pods was found in T1. The treatment pairs T2, T3 and T2, T4 were significantly different but the pair T3, T4 may be regarded as similar in their performance.

5. CONCLUSION

The use of banana plant pseudo-stem juice is significantly effective in the cultivation of mustard. Use of the juice in lieu of muriate of potash affords 23% higher yield of mustard as compared to that with muriate of potash. Using banana plant juice, one can achieve up to 65% higher yield over normal soil. It is observed that both the average number of pods per plant as well as the average weight of seeds are higher with banana plant pseudo-stem juice as compared to corresponding values with muriate of potash. Thus, banana plant pseudo-stem juice is a better replacement for muriate of potash in the cultivation of mustard.

REFERENCES

[1]   *Annual Report 2018-19*; National Commodity & Derivatives Exchange Limited: Mumbai (India), **2019**.

[2]   *Mustard oil Market - Global Industry Analysis, Size, Share, Growth, Trends, and Forecast 2017 – 2027*, **2017**. https://www.transparencymarketresearch.com/mustard-oil-market.html

[3]   Sahu, M.; Devi, S.; Mishra, P.; Gupta, E. Mustard is a miracle seed to human health. In: *Ethnopharmacological investigation of Indian spices*, 1st ed; Mishra, N., Ed.; NIGI Global Publisher (online), **2020**; pp. 154-162. Ch 12.

CHAPTER 6

Organic Potash-rich Banana Plant Pseudo-stem: Application in Chili Cultivation

Abstract: Banana plant pseudo-stem juice is rich in potassium, which is present in organic form. Its use in chili cultivation has been discussed. If banana plant pseudo-stem juice is used in lieu of muriate of potash, 20% more yield is achieved. As compared to potash treated soil, juice treated soil affords a higher number of chilies per plant, a higher average per capita weight of chili as well as a higher total yield per unit of land. Overall yield in juice treated soil is 16-fold higher as compared to that in normal soil and 5-fold higher as compared to that in soil treated with urea and super phosphate fertilizer but not with potash. Thus, organic potash present in banana plant pseudo-stem juice is a better replacement for commercial potash fertilizer.

Keywords: Alternative of MOP, Chili cultivation, Organic potash, Use of banana plant pseudo-stem juice.

1. INTRODUCTION

Chili is the most widely used universal spice and is produced by several countries across the globe. India, China, Thailand, Pakistan, Myanmar, Bangladesh, Vietnam, Mexico, Nigeria, Romania, Japan, Ethiopia, Uganda, Turkey and Indonesia are the major chili producing countries, and its global production is around 7 million tones [1]. Chili adds pungency, taste, flavor as well as color to dishes, and therefore is considered an important ingredient in cuisines around the world. Because of the color and pungency, Indian red chili has high demand in the global market. Among Indian spices, chili is the major export commodity.

Chilies are used as both green and ripe; also, red chilies are available in dry and powdered form. Fresh green chilies are considered to have a number of health benefits. They are rich in vitamins, good antioxidants and speed up digestive metabolism [2]. Chilies stimulate taste buds in the mouth and thereby increase the secretion of enzyme-loaded saliva. Saliva is rich in amylase which helps in the breakdown of foods. Pungent chilies generally find use as a flavoring agent in cooked meals. The non-pungent varieties are cooked as vegetables or used with other food items for flavor. Strongly pungent varieties are used as a spice for seasoning and are generally consumed in small amounts to stimulate appetite. Chilies, along with salt and vinegar, are also used to prepare pickle. Apart from its extensive use in spices, chili varieties are also used in food processing industries, such as a coloring agent in salad dressings, meat products, cosmetics, and even clothing [3]. Chili is rich in capsaicin which finds applications in pharmaceutical preparations for cold, sore throat, chest congestion, *etc*. It is also used in skin ointments and powders. Oleoresin with ingredient input from chili is an important cosmetic product having high demand in European countries [4].

Across the globe, nearly 400 different varieties of chilies are reported. All these varieties belong to the family Solanaceae and genus *Capsicum*. Three important species of *Capsicum* namely, *C. annuum, C. frutescens* and *C. chinense* evolved from a common ancestor that grew wildly in the North of the Amazon basin (NW-Brazil, Columbia). In the course of time, these species were domesticated, and many other species were developed through the process of hybridization, naturally or artificially. In Guinness Book of World Records 2013, 'Carolina Reaper' is the hottest variety of chili, having a pungency of about 2.2 million SHU (Scoville Heat Units) and it was developed by a grower Ed Currie of West Indies [5]. Other varieties of chili certified earlier as hottest are Infinity chili in 2011 and 'Bhut Jolokia' or Naga Jolokia' in 2007. 'Bhut Jolokia' or 'Naga Jolokia' is a native of North-East India and reported as a hybrid of *Capsicum chinense* Jacq. and *Capsicum frutescens* L. [6]. 'Bhut Jolokia' is one of the crops that we have used in our experiments on the use of banana plant juice as a substitute for commercial potash.

2. EXPERIMENTAL DETAILS

2.1. Materials

'Bhut Jolokia' or 'Naga Jolokia' was cultivated. For more details, Section 2.1 of Chapter 3 may be consulted.

2.2. Details of Methodology

Cultivation of chili was carried out in earthen pots (1 ft diameter and 1 ft depth) and the soil in the pots was given four different treatments. Eight pots (two pots for each treatment) were used. Soil was prepared based on the soil testing report (*cf.* Table **1** in Chapter 3) and recommended quantities of fertilizers for chili cultivation (Table **1**) from the Department of Agriculture, Lakhimpur district, Assam. To provide potassium to soil in treatment T2, banana plant juice was applied, and in treatment T3, commercial fertilizer muriate of potash (MOP) was applied. No potash was applied in the treatment T1 and T_4. Equal quantities of urea to provide nitrogen and single super phosphate (SSP) to provide phosphorus were applied in the treatments T2, T3 and T4. Neither chemical fertilizer nor banana plant juice was used in treatment T1. It was used as the control. Different treatments are shown in Table **2**.

Table 1. Fertilizers recommended for chili cultivation (in kg/bigha, 7.47 bigha = 1 hectare).

Farm Yard Manure (FYM)	= 1500 kg
Urea	= 32.0 kg
Single Super Phosphate (SSP)	= 26.0 kg
Muriate of Potash (MOP)	= 12.0 kg
Lime	= not required

Table 2. Fertilizers applied in different pots for chili cultivation.

Plot	Treatments	Urea (g/pot)	SSP (g/pot)	MOP (g/pot)	Banana Plant Juice (Litre/Pot)
1	T1	Nil	Nil	Nil	Nil
2	T2	2.3	1.8	Nil	0.230
3	T3	2.3	1.8	1	Nil
4	T4	2.3	1.8	Nil	Nil

Procedure for plantation: All the eight earthen pots were of equal size and shape (1 ft diameter and 1 ft depth). Bottoms were perforated to allow the excess water to flow out. About one inch of each pot was filled with a mixture of fine stones and coarse sand. The requisite quantity of soil from normal agricultural land, which was already tested, was collected, finely powdered and uniformly mixed with FYM in a 5:1 ratio (Soil: FYM). The two pots used for the treatment T1 were filled with the mixture. Calculated quantities of urea and SSP (Table **3**) were applied to the remaining part of the soil mixture and the pots for treatments T2 and T4 were filled up. The last part of the soil mixture was then mixed with calculated quantities of MOP and used to fill up the two pots for treatment T3. In treatment T2, the soil in the pots was treated with banana plant juice (230 mL per pot, Table **2**) in lieu of MOP. The soil-filled pots were covered with polythene and left untouched under the shed for 3 days. After 3 days, healthy chili seedlings (one in each pot) were planted and tap water was added to moisten the soil around the seedlings. The pots were kept in shade for the next 5 days to protect the seedlings from dehydration and to facilitate new roots to grow. Every evening, water was added to pots to moisten the soil appropriately but not to make it soggy. After 5 days, the pots were kept in an open environment and tap water was added time to time to keep the soil appropriately moistened but not soggy. The plantation was done in the 1st week of September.

Procedure for the collection of matured chilies: Yields in different treatments were estimated by collecting ripe chilies periodically from each treatment and recording their weights. Ripe chilies from different treatments at different time intervals were collected, their numbers counted and total weights recorded (Table 3). Harvesting was started from the 1st week of January.

3. RESULTS AND DISCUSSION

The methods for the estimation of macro and micronutrients in banana plant pseudo-stem juice are described in Chapter 2 and results are discussed in Chapter 3 (Section 3).

Most of the macro and micronutrients are usually available in the soil. However, macronutrients such as potassium, nitrogen and phosphorus get quickly depleted because plants use them substantially. Therefore, the soil needs replenishment of macronutrients. Usually, urea is used to enrich the soil with nitrogen, SSP to enrich the soil with phosphorus and MOP to enrich the soil with potassium. Here, in our experiment, we have tried to evaluate the efficacy of banana plant pseudo-stem juice as a substitute for MOP. The presence of a considerable amount of K^+ in the banana pseudo-stem juice is noteworthy.

Photographs of chili cultivation are shown in Figs. (**1** to **5**). As compared to saplings in treatments T1 and T4, saplings in treatments T2 and T3 are developing faster and healthier. Also, saplings in treatments T2 and T3 are greener as compared to those in treatments T1 and T4. Better development of saplings in treatment T2 and T3 indicates that potassium along with nitrogen and phosphorus is essential for plants to maintain good health. Also, we can observe that effect of banana plant pseudo-stem juice is as good as MOP if not better. The slower development of saplings in T1 and T4 is due to the absence of adequate quantities of potassium, nitrogen and phosphorus in soil. Relatively better development of saplings in treatment T4 as compared to treatment T1 is due to the effect of urea and SSP. While urea and SSP have been applied in treatment T4, none is applied in treatment T1.

Fig. (1). After 30 days of plantation.

Fig. (2). After 60 days of plantation.

Fig. (3). After 70 days of plantation.

Fig. (4). After 80 days of plantation.

Yields from different treatments are recorded in terms of the number of chilies at different time intervals (Table **3**), total weight of chilies collected periodically (Table **4**) and the average per capita weight of chilies at different time intervals (Table **5**). In the first batch of harvesting after 120 days, 14 pieces of chilies were collected from the treatment T2 (banana plant juice treatment) against only 3 pieces from treatment T1 (control, normal soil), 9 pieces from treatment T3 (chemical potash treatment) and 7 pieces from treatment T4 (without potash). These results show that there is nearly a 5-fold increase in the number of chilies in the treatment with banana plant juice as compared to normal soil, more than 1.5-fold increase as compared to treatment with chemical potash and a 2-fold increase as compared to treatment without potash. Results also indicate that all the three nitrogen, phosphorus and potassium fertilizers together perform better, and banana plant juice as the source of potassium is superior to MOP and the performance of nitrogen-phosphorus combination without potassium fertilizer is poor.

Fig. (5). After 100 days of plantation.

Table 3. Number of ripe chilies from different treatments.

No. of Days	Number of Ripe Chilies from Different Treatments			
	T1	T2	T3	T4
120	3	14	9	7
127	2	10	15	-
139	1	12	21	5
166	-	21	19	2
185	-	18	3	1

The total weight of chilies from treatments at different time intervals is shown in Table **4**. The weight of the 1ˢᵗ batch of chilies from treatment T2 is nearly 7.5-fold higher as compared to that in T1, nearly 1.5-fold higher as compared to that in T3 and more than 2-fold higher as compared to that in T4. After the first batch of crop harvesting, the yield from treatment T1 started falling, and in 3 weeks, it completely stopped. On the other hand, in soil treated with banana plant juice (treatment T2), the plants remained productive for another two months almost at an equal level. In treatment with MOP, the plants continued to yield fruits almost at an equal level of productivity for another one and half months and then yields started falling drastically. In treatment T4, the plants continued to give yield for another one and half months but at much lower productivity. Thus, banana plant juice not only enhances the productivity of chili plants but also enhances their productive life.

Table 4. Total weight of ripe chilies from different treatments.

No. of Days	Total Weight of Ripe Chilies Per Treatment			
	T1	T2	T3	T4
120	7.5	55.9	38.0	26.6
127	6.1	40.3	52.5	-
139	2.5	43.2	63.0	16.2
166	-	67.2	58.9	6.4
185	-	59.4	8.8	2.5

Per capita weight of chilies in treatments T2 and T3 is almost equal, thus indicating the importance of potassium in maintaining the size and weight of chilies (Table **5**). The per capita weight of chilies in treatment T4 is lower and in treatment T1 much lower as compared to that in T2 and T3. In treatment T4, the presence of nitrogen and phosphorus fertilizers has helped in better utilization of whatever amount of potassium was present in soil and therefore initial per capita weight of chilies in T4 was better but started falling thereafter as the potassium availability depleted. On the other hand, per capita weight of chilies in T1 is much lower from the very first batch of chilies because of lack of adequate quantities of macronutrients such as nitrogen, phosphorus and potassium in normal agricultural soil.

Table 5. Average weight per chili from different treatments.

Date	Average Weight Per Chili from Different Treatments (g)			
	T1	T2	T3	T4
120	2.5	4.0	4.2	3.8
127	3.0	4.0	3.5	-
139	2.5	3.6	3.0	3.2
166	-	3.2	3.1	3.2
185	-	3.3	2.9	2.5

4. STATISTICAL ANALYSIS

The data collected were statistically analyzed. Table **6** shows that the number of chilies is significantly varied (P<0.05) among the treatments. The higher number of chili is recorded in the treatment T2 where banana pseudo-stem juice was applied, followed by T3, T4 and T1. The average number of chilies in different treatments does show that treatments T2, T3 and T4 are significantly different from the control T1. The treatment pairs T2, T4 and T3, T4 are significantly different from each other but the treatment pair T2, T3 may be regarded as similar in their performance.

Table 6. Statistical Analysis.

Entry	Treatment	Average Number of Chilies	Average Weight Per Chilli (g)	Total Yield (g)
1	T1	1.20 ± 0.43^c	1.80 ± 0.48^a	16.1
2	T2	15.00 ± 1.78^a	3.55 ± 0.15^b	266.3
3	T3	13.40 ± 2.96^a	3.30 ± 0.21^b	221.2
4	T4	3.10 ± 0.67^b	2.64 ± 0.51^b	51.9
	F Test	*	*	

Values represent mean ± standard error of five replicates
F test: *: P < 0.05
a, b, c and d: Means followed by the same letter are not significantly different according to Tukey's method at 5% level.

Table **6** also shows that the average weight of chilies is significantly different (P< 0.05) among treatments. The critical difference value and the average weight of chili under different treatments reveal that treatments T2, T3 and T4 are significantly different from the control, T1. Even though there is no significant difference among the treatments T2, T3 and T4, the average weight of chilies is found highest (3.55 g) in the treatment T2 where banana pseudo-stem juice was applied in lieu of potash. This may be the benefits of other micronutrients present in banana plant juice. The higher number and higher average weight of chilies in treatment T_2 together account for the higher total yield of chili in the treatment T_2. Thus, banana pseudo-stem juice can be used as an easily affordable organic fertilizer to replace commercial MOP. The total yields of chili in different treatments are graphically shown in Fig. (**6**).

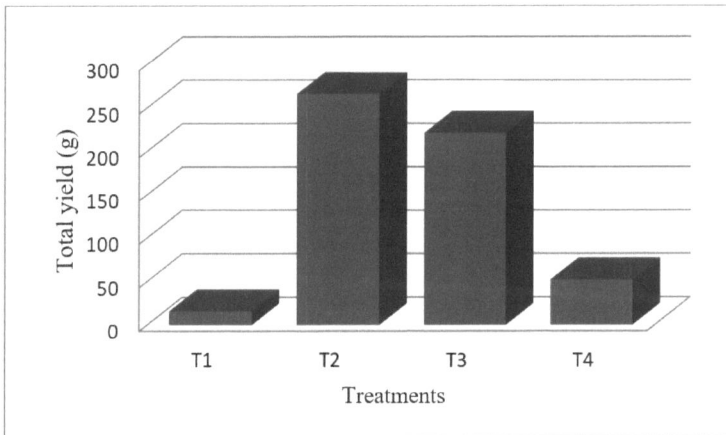

Fig. (6). Total yield of chili in different treatments.

5. CONCLUSION

In conclusion, the juice of banana plant pseudo-stem is found significantly effective in the cultivation of chili. Application of banana plant pseudo-stem juice in lieu of muriate of potash affords 20% higher yield of chilies, more than 16-fold increase in yield (1654% enhancement) as compared to that in normal soil, and more than 5-fold increase (513% enhancement) as compared to that in soil where nitrogen and phosphorus fertilizers are applied but not potassium fertilizer. When banana plant pseudo-stem juice is applied in lieu of chemical potash, both the average number of chilies per plant as well as the average per capita weight of chilies are improved. Thus, banana plant pseudo-stem juice is a better alternative to MOP in the cultivation of chili.

REFERENCES

[1]   a) Geetha, R.; Selvarani, K. A study of chilli production and export from India. *IJARIIE,* **2017**, *3*(2), 205-210. b) Chilli Outlook. *Agricultural Market Intelligence Centre*; Professor Jayashankar Telangana State Agricultural University: Rajendranagar, Hyderabad (India), **2019**, pp. 1-4.

[2]   a) McGee, H. On food and cooking: the science and lore of the kitchen. *Completely Revised and Updated,* **2004**, , 418-421. Ch 8. b) Peppers may increase energy expenditure in people trying to lose weight. *Sci. News,* **2010**. April 28.

[3]   a) Saxena, A.; Raghuwanshi, R.; Gupta, V.K.; Singh, H.B. Chilli anthracnose: The epidemiology and management. *Front. Microbiol.,* **2016**, *7*, 1527.
       [http://dx.doi.org/10.3389/fmicb.2016.01527] [PMID: 27746765]. b)Dagnoko, S.; Yaro-Diarisso, N.; Sanogo, P.N.; Adetula, O.; Dolo-Nantoumé, A.; Gamby-Touré, K.; Traoré-Théra, A.; Katilé, S.; Diallo-Ba, D. Overview of pepper (*Capsicum* spp.) breeding in West Africa. *Afr. J. Agric. Res.,* **2013**, *8*(13), 1108-1114.
       [http://dx.doi.org/10.5897/AJAR2012.1758]

[4]   *Botany and Economic Importance of Chilli,* **2012**, https://www.agropedia.iitk.ac.in/content/botany-and-economic-importance-chilli online resource material

[5]   *PuckerButt Pepper Company claims Guinness World Record Title for Hottest Chili,* **2013**, www.puckerbuttpeppercompany.com online

[6]   a) Bosland, P.W.; Baral, J.B. 'Bhut Jolokia'-The world's hottest known chile pepper is a putative naturally occurring interspecific hybrid. *Hort Science,* **2007**, *42*(2), 222-224.
       [http://dx.doi.org/10.21273/HORTSCI.42.2.222]. b)Meetei, N.T.; Singh, A.K.; Singh, B.K.; Mandal, N. Recent advances in naga king chilli (*capsicum chinense* JACQ.) research. *Int. J. Agric. Environ. Biotechnol.,* **2016**, *9*(3), 421-428.
       [http://dx.doi.org/10.5958/2230-732X.2016.00054.1]

CHAPTER 7

Potash-rich Banana Plant Pseudo-stem Juice: Application in Brinjal (Eggplant) Cultivation

Abstract: The use of potash-rich banana plant pseudo-stem juice in the cultivation of brinjal (eggplant) has been discussed. The yield of brinjal (aubergine) from the soil treated with banana plant pseudo-stem juice is significantly higher as compared to that from the soil treated with muriate of potash. Juice treated soil shows a higher number of brinjals per plant, higher average per capita weight as well as higher yield per unit of land as compared to potash treated soil. The overall yield from juice treated soil is 34% more as compared to that from potash treated soil; nearly 200% more than that from soil not treated with potash. Thus, the banana plant pseudo-stem juice is a better replacement for potash fertilizer. Colour pictures are reported

Keywords: Aubergine cultivation, Eggplant crop, Organic potash, Potash for brinjal cultivation, Use of banana plant juice, Use of banana plant pseudo-stem.

1. INTRODUCTION

Brinjal or eggplant, or aubergine, is a major vegetable in several countries across the globe. While the name brinjal is popular in South Asia and South African countries, the name eggplant is popular in the USA, Australia, New Zealand, and Canada. It is popularly known as aubergine in European countries, especially in France. The botanical name of brinjal is *Solanum melongena* L., family Solanaceae, genus Solanum. The species *Solanum melongena* has a large number of cultivars. Considered native to India, it is an annual herbaceous plant in tropics and subtropics and is widely grown in Asian countries [1]. The estimated global annual production of brinjal is more than 50 million tons. After potato, tomato, pepper, and tobacco, it is considered the fifth most economically important solanaceous crop. The top five brinjal producing countries are China, India, Egypt, Turkey, and Iran [2].

Dibakar Chandra Deka & Satya Ranjan Neog

Brinjal contains mostly water, some protein, and fiber. It is low in carbohydrates and fats. It is a good source of minerals, vitamins, and antioxidants [3]. In traditional medicine, brinjal is believed to have several medicinal values against a number of common ailments. In Ayurved, it is recommended for those suffering from liver complaints and diabetes [4].

Brinjal is usually transplanted rather than directly seeded in the field. Transplantation facilitates uniform and adequate space between plants. Brinjal can be grown on soils of a wide range of textures, from light sandy soil to silt loamy to heavy clay soil. It can tolerate moderate acidity. However, the ideal soil is well-drained loamy silt and loamy clay soil with a pH range of 5.5-6.8 [5]. Soil with a high level of pH should be treated with farmyard manure before transplantation. Brinjal is susceptible to severe frost and requires a long warm growing season for successful cultivation. The optimum temperature is 20-30 °C [4]. In the North-East of India, brinjal can be grown around the year, the main planting season being October to November in the plains and July to August in the hills.

2. EXPERIMENTAL DETAILS

2.1. Materials

The brinjal variety which was cultivated is not scientifically identified. It was a domesticated variety bearing cylindrical fruits that may grow about 12 inches long with a diameter up to 2 inches. For other details, Section 2.1 of Chapter 3 may be consulted.

2.2. Details of Methodology

Before transplantation of brinjal saplings, the soil was prepared following the guidelines from the Department of Agriculture, Lakhimpur district, Assam. The soil was tested, and based on the test report (*cf.* Table **1** in Chapter 3), quantities of fertilizers were recommended (Table **1**). The cultivation of brinjal was carried out in pots. The soil used to fill the pots was given four different treatments. Total eight pots, two for each treatment, each pot of 1 ft depth and 1 ft diameter, were used.

Quantities of fertilizers recommended for brinjal cultivation are shown in Tables **1**, and **2**, quantities of fertilizers used for each pot in four different treatments, T1, T2, T3, and T4, are recorded. Banana plant (variety *Musa balbisiana*) juice as a source of potassium was applied to the soil of treatment T2, whereas chemical fertilizer muriate of potash (MOP) was used in treatment T3. No potash was applied in the treatments T1 and T4. Equal quantities of urea and SSP were applied in the treatments T2, T3, and T4. Treatment T1 was used as the control, neither chemical fertilizers nor banana plant juice was applied.

Table 1. Fertilizers recommended for brinjal cultivation in kg/bigha (7.47 bigha = 1 hectare).

Farm Yard Manure (FYM)	= 1500 kg
Urea	= 23.0 kg
Single Super Phosphate (SSP)	= 31.0 kg
Muriate of Potash (MOP)	= 10.0 kg
Lime	= not required

Table 2. Fertilizers applied to soil in different pots for brinjal cultivation.

Pots	Treatments	Urea (g)	SSP (g)	MOP (g)	Banana Plant Juice (Litres)
1	T1	Nil	Nil	Nil	Nil
2	T2	1.6	2.2	Nil	0.160
3	T3	1.6	2.2	0.7	Nil
4	T4	1.6	2.2	Nil	Nil

Procedure for plantation: Brinjal cultivation was done in pots, each of 1 ft depth and 1 ft diameter. Soil from normal agricultural land, after appropriate treatments, was used to fill in the pots. An adequate quantity of soil was collected, appropriately treated for different treatments, namely T1, T2, T3, and T4, with quantities of fertilizers shown in Table **2**, and used to fill in the pots. Altogether, eight pots, two for each treatment, were used.

The pots were perforated at the bottom to allow excess water to drain out. About 1 inch from the bottom of each pot was filled with a mixture of sand and small gravels. The soil for all treatments was uniformly mixed with FYM in a 5:1 ratio (Soil: FYM). The pots for the treatment T1 were filled with the mixture, and no other fertilizer was used. The soil mixtures for treatments T2, T3, and T4 were uniformly mixed with the calculated quantities of urea and SSP (quantities shown in Table **2**), and the pots for treatment T4 were filled with. The soil mixture for treatment T3 was mixed with the calculated quantity of MOP, and to the soil mixture for T2, in lieu of MOP, banana plant juice (160 mL per pot, Table **2**) was added. The soil-filled pots were covered with polythene and left untouched under a shed for 72 h. On the 4th day, healthy and carefully selected brinjal saplings were transplanted, one in each pot, and enough quantity of tap water was added to moisten the soils around the saplings. The pots were kept in shades for the next 5 days to protect the saplings from dehydration and to facilitate new roots to grow. Water was added to the pots from time to time to make the soil appropriately moistened but not soggy. After 5 days, the pots were kept in an open environment, and enough tap water was poured from time to time to keep the soil appropriated moistened but not soggy. The plantation was carried out in the early part of the month of January.

Procedure for brinjal collection: Brinjals were collected at the edible stage. The collection was planned in such a way that number of brinjals at different time intervals, and their weights in different treatments can be compared. Numbers of brinjals from different treatments at different time intervals are shown in Table **3**, the average weight per brinjal in different treatments in Table **4**, and total yield per plant in different treatments in Table **5**. Harvesting was started from the 4th week of March (after nearly 3 months of the plantation).

3. RESULTS AND DISCUSSION

Nutrients, both micro and macro, present in banana plant pseudo-stem juice were estimated by methods described in Chapter 2, and results discussed in Section 3, Chapter 3.

Plants depend on soil for most of their micro- and macronutrients. As macronutrients are absorbed in large quantities, the soil gets quickly depleted, and therefore, the soil needs periodical replenishment with macronutrients. Nitrogen, phosphorus, and potassium are three essential macronutrients which plant acquire from the soil. Here in our experiments, we replenished soil with SSP to provide phosphorus and urea to replenish the soil with nitrogen, but to replenish the soil with potassium, banana plant pseudo-stem juice was used in lieu of MOP.

Photographs of brinjal cultivation in four different treatments are shown in Figs. (**1** to **5**). In treatments T2 and T3, saplings developed faster, also looked healthier and greener as compared to those in treatments T1 and T4. To provide potassium, banana plant juice (BPJ) was applied in treatment T2, while MOP was applied in treatment T3. The photographs provide a clear indication that BPJ, as the source of potassium, is as good as MOP, even better. In treatments T1 and T4, no source of potassium, neither BPJ nor MOP was applied, and therefore the growth of saplings was slower due to lack of adequate supply of potassium from the soil. Relatively better growth of saplings in treatment T4 as compared to T1 is due to the effect of either urea or SSP or both. None of these two fertilizers was applied to soil in treatment T1.

Fig. (1). Brinjal saplings after 25 days of plantation.

Fig. (2). Brinjal saplings after 60 days.

Fig. (3). Brinjal saplings after 83 days.

The productivity of plants in different treatments is compared in terms of the number of brinjals (Table **3**), the average weight per brinjal (Table **4**), and total yield (Table **5**). Table **3** shows the numbers of brinjals collected from different treatments at different time intervals. In the first batch of harvest after 83 days, 5 pieces of brinjals were collected from the treatment T2 (with BPJ) against zero pieces from the treatment T1 (normal soil, control), 1 piece from the treatment T3 (with MOP), and none from the treatment T4 (without potash), thus indicating the possibility of early harvesting as well as higher productivity with banana plant juice treatment (treatment T2) as compared to normal soil (treatment T1) and soil without potash (treatment T4). When the soil was treated with banana plant juice (treatment T2), a 5-fold increase in productivity was observed compared to the treatment with MOP (treatment T3). Within 31 days of the 1st harvest, a total of 20 pieces of brinjal from the treatment T2 against 1 piece from T1, 11 pieces from T3, and 4 pieces from T4 were collected. The results indicate that the productivity of brinjal is much higher, almost double, when banana plant juice is used as the substitute for MOP. Yields in treatments T1 and T4 indicate that the productivity of brinjal cultivation is not only delayed but also much lower without potash. Within 45 days of harvest, a total of 34 pieces of brinjal were collected from the treatment T2 against only 4 pieces from T1, 28 pieces from T3, and 15 pieces from T4. After 45 days of harvesting, the plants in treatments either with banana plant juice or MOP continued to yield for a couple of days more but at reduced productivity. Plants on normal soil (treatment T1) completely stopped yielding fruits after 45 days, and plants on soil treated with nitrogen and phosphorus fertilizers but without potassium fertilizer continued to yield for a few days more. Still, it then stopped much earlier than treatments T2 and T3. From these observations, it can be concluded that potash is important for brinjal cultivation; it enhances not only productivity but also enables farmers to have early harvesting and for a longer duration. The delayed harvesting and shorter duration of harvest in treatments T1 and T4 are significant observations. The results indicate that the delay in productivity and the shorter duration of productive life is probably due to an inadequate supply of potash.

Fig. (4). Brinjal saplings after 96 days.

The average per capita weight of brinjals at different time intervals in treatment T2 is consistently higher than that in T3, much higher than those in T1 and T4 (Table **4**). The per capita weight of brinjals in T3 is also higher than those in treatments T1 and T4. Thus, all these observations indicate that potash helps not only early harvesting and to achieve a higher number of brinjals but also better crops in terms of size and weight of brinjals. It is interesting that replacing MOP with banana plant juice affords better crops in all respects, say the size, number, or per capita weight of brinjals; also, plants continue to give yields for a longer duration. It is, therefore, clear that banana plant juice is a better substitute for MOP in the cultivation of brinjal.

Fig. (5). Brinjal saplings after 126 days.

Table 3. Number of brinjals from different treatments at different time intervals.

No. of Days	Number of Brinjals from Different Treatments			
	T1	T2	T3	T4
83	-	5	1	-
90	-	7	3	1
114	1	8	7	3
119	2	7	9	5
127	1	7	8	6

Table 4. Average weight per brinjal in different treatments.

No. of days	Average Weight of Brinjals (g)			
	T1	T2	T3	T4
83	-	92.8	70.0	-
90	-	89.1	76.7	62.0
114	48.0	85.0	80.6	68.0
119	54.0	85.4	75.6	62.8
127	52.0	80.0	79.3	66.3

Table **5** shows the number of brinjals per plant, per capita weight of brinjals, and total productivity per plant in 127 days of the plantation. The number of brinjals from treatment T2 is 8.5-fold higher as compared to that on normal soil (treatment T1), 1.2-fold higher as compared to that on soil treated with MOP (treatment T3), and more than 2.2-fold higher as compared to that on soil treated with urea and phosphate fertilizers but without potassium. If the average per capita weight of brinjals from soils with different treatments are compared, it is observed that brinjals from soil treated with banana plant juice weigh 65% more than that from normal soil, 10% more than that from soil treated with MOP, and 32% higher than that from soil without potash. Total productivity per plant is 14-fold increased over normal soil when the soil is treated with banana plant juice along with adequate nitrogen and phosphorus. Productivity per plant is 34% enhanced when MOP is replaced by banana plant juice. It is interesting to note that the addition of potash to soil with adequate nitrogen and phosphorus supplements enhances the productivity of plants by 122% (treatments T3 *vs.* T4). When commercial potash is replaced with banana plant juice, the productivity of plants is further enhanced to 199% (treatments T2 *vs.* T3). Thus, banana plant pseudo-stem juice is an ideal greener substitute for MOP in the cultivation of brinjal.

Table 5. Average number of brinjals per plant, the average per capita weight and total yield per plant in 127 days from the plantation.

Treatments	Average Number of Brinjals Per Plant	Average Weight Per Brinjal (g)	Productivity Per Plant (g)
T1	2	52	104
T2	17	86.06	1463
T3	14	77.79	1089
T4	7.5	65.20	489

4. STATISTICAL ANALYSIS

Total yields by weight for the whole duration from different treatments, average per capita weight of brinjals based on the total numbers, and the average number of brinjals per plant in different treatments are shown in Table **6**. To understand the significance of different treatments, all data are statistically analyzed by analysis of variance (using ANOVA table), and the mean separations are done using Tukey's method at a 5% level.

Table 6. Statistical Analysis.

Entry	Treatments	Average Number of Brinjals	Average Weight Per Brinjal (g)	Total Yield (g)
1	T1	1.0 ± 0.25^c	31.00 ± 11.17^d	208
2	T2	6.8 ± 0.44^a	86.46 ± 1.92^a	2926
3	T3	5.6 ± 1.37^{ab}	76.44 ± 1.65^b	2178
4	T4	3.1 ± 0.96^{bc}	51.92 ± 11.54^c	978
	F Test	*	*	

Value represents mean ± standard error of five replicates
F test: *: $P < 0.05$,
a, b, c and d: Means followed by the same letter are not significantly different.

Table **6** shows the statistical analysis of the average number of brinjals per plant, average weight of brinjals, and total yield per treatment. Statistical analysis on the number of brinjals per plant shows that there exists a significant difference among different treatments. Still, there is no significant difference between the treatment with muriate of potash (treatment T3) and the treatment with banana plant juice (treatment T2). However, treatment T2 shows a higher number of brinjals than the treatment T3. This is noteworthy and should be considered an additional bonus for the use of banana plant juice in lieu of MOP. There is no significant difference between treatment pairs (T1, T4) and (T3, T4), but the treatments T2 and T3 are significantly different from treatment T1. Among the treatments, the highest number of brinjals is produced in the treatment T2.

The average weight of brinjals is significantly varied among the treatments. The highest average weight of brinjal is recorded in the treatment where banana plant juice was applied (treatment T2), followed by treatments T3, T4, and T1. The critical difference value shows that there exists a significant difference between the treatment pairs (T1, T4), (T4, T3), and (T3, T2), but the highest average weight of brinjals is achieved in the treatment T2; this may be due to the presence of not only potassium but also other nutrients in banana plant juice. The highest yield is obtained in the treatment T2 followed by treatments T3, T4, and T1.

The results discussed above are graphically presented below. Figs (**6** - **8**) show, respectively, the number of brinjals per plant, average per capita weight of brinjals, and total yield of brinjals from different treatments.

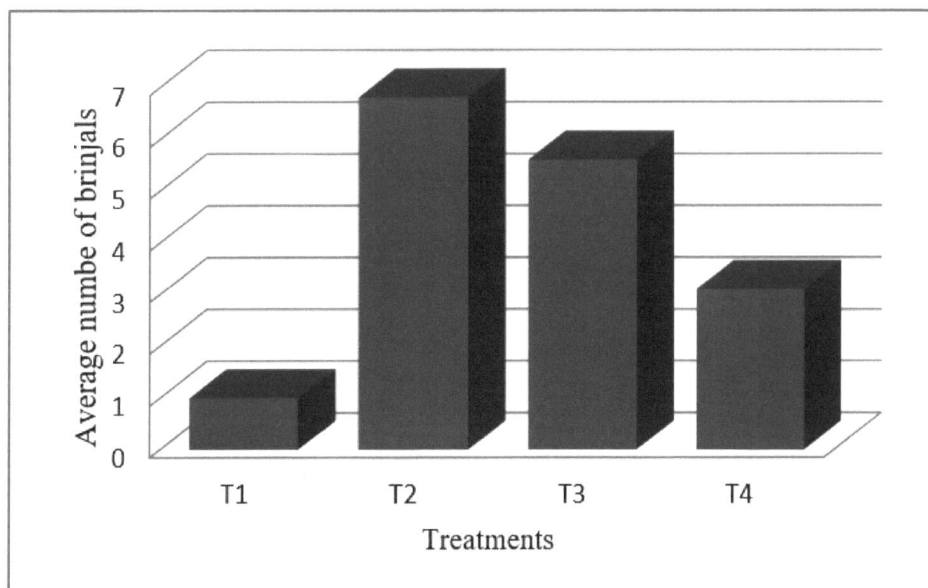

Fig. (6). Average number of brinjals per plant in different treatments.

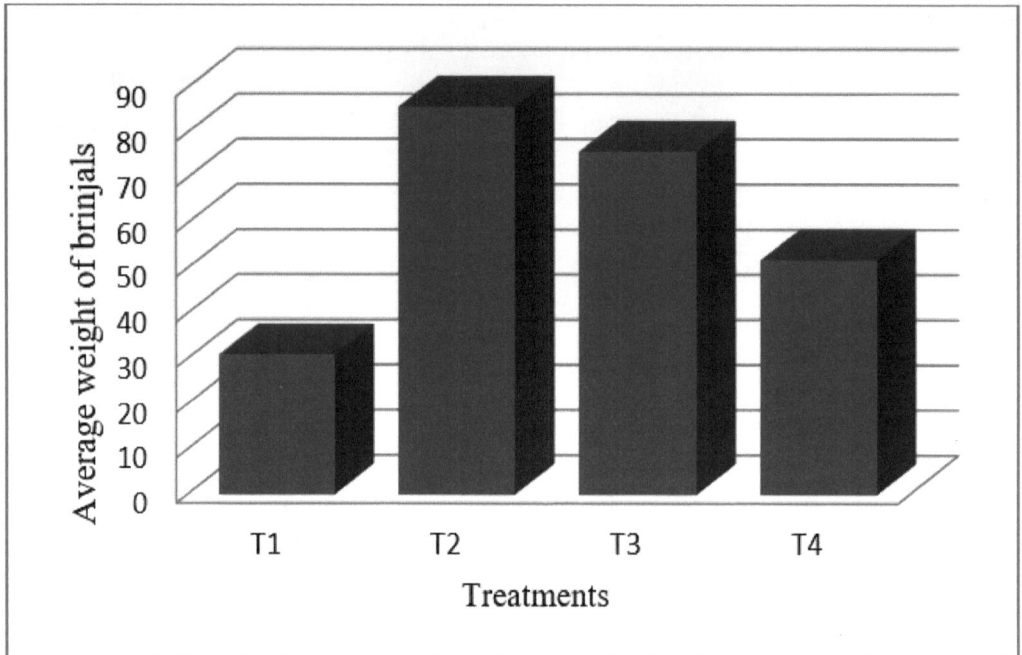

Fig. (7). Average weight of brinjals in different treatments (g).

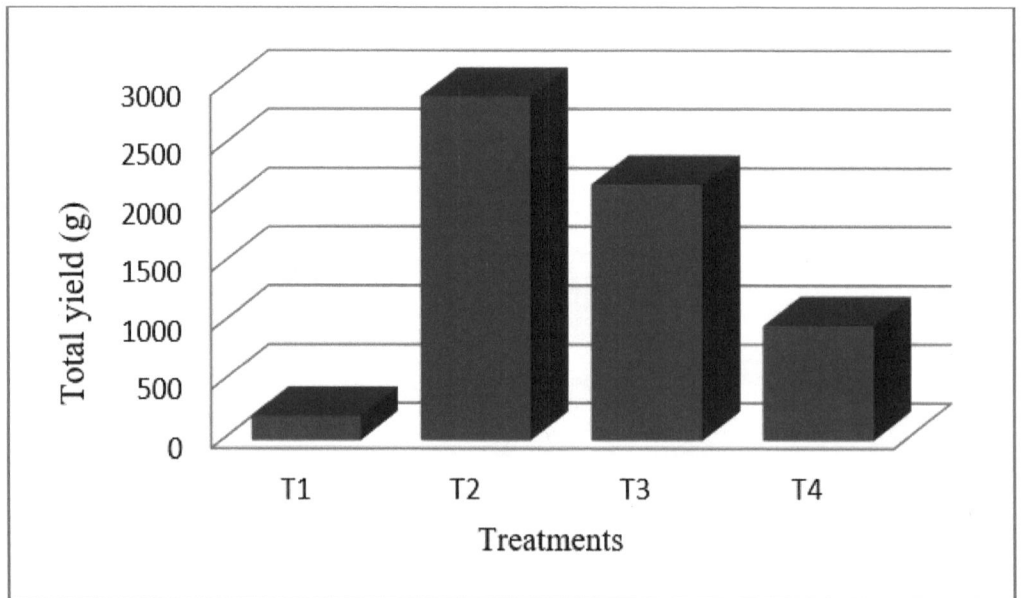

Fig. (8). Total yield of brinjals in different treatments (g).

5. CONCLUSION

The juice of the banana plant pseudo-stem is rich in potassium and some other micronutrients. Its application in brinjal cultivation is significantly effective. Application of banana plant pseudo-stem juice as potash supplement affords 34% higher yield of brinjals over that with commercial MOP, more than 14-fold increase (1406% enhancement) as compared to yield from normal soil and nearly 3-fold increase (299% enhancement) as compared to yield in soil with nitrogen and phosphorus fertilizers but without potassium fertilizer. It is observed that both the average number of brinjals per plant and the average per capita weight of brinjals are higher in soil treated with banana plant pseudo-stem juice as compared to corresponding values in soil treated with MOP. Also, 222% productivity enhancement is observed when the adequate quantity of MOP as a potassium supplement along with urea and SSP fertilizers are added to normal agricultural soil. If banana plant juice is added in lieu of MOP along with urea and SSP, productivity enhancement is 299%, an additional 77% bonus because of banana plant juice. Thus, banana plant pseudo-stem juice is an ideal and better replacement for MOP in the cultivation of brinjal.

REFERENCES

[1]   *Eggplant, Wikipedia the Free Enclopedia,* **2012**. accessed 2020-07-01

[2]   a) *FAO Statistical Yearbook 2012,* **2013**. b) Taher, D.; Solberg, S.Ø.; Prohens, J.; Chou, Y-Y.; Rakha, M.; Wu, T.H. World Vegetable center eggplant collection: Origin, composition, seed dissemination and utilization in breeding. *Front. Plant Sci.,* **2017**, *8*, 1484.
       [http://dx.doi.org/10.3389/fpls.2017.01484] [PMID: 28970840]

[3]   a) Raigón, M.D.; Prohens, J. Muñoz-Falcón, J.E.; Nuez, F. (2008). Comparison of eggplant landraces and commercial varieties for fruit content of phenolics, minerals, dry matter and protein. *J. Food Compos. Anal.,* **2008**, *21*, 370-376.
       [http://dx.doi.org/10.1016/j.jfca.2008.03.006]. b) Prohens, J.; Plazas, M.; Raigón, M.D.; Seguí-Simarro, J.M.; Stommel, J.R.; Vilanova, S. Characterization of interspecific hybrids and first backcross generations from crosses between two cultivated eggplants (*Solanum melongena* and *S. aethiopicum* Kumba group) and implications for eggplant breeding. *Euphytica,* **2012**, *186*, 517-538.
       [http://dx.doi.org/10.1007/s10681-012-0652-x]

[4]   *Biology of Brinjal; Department of Biotechnology*; Ministry of Science & Technology (Govt. of India): New Delhi (India), **2012**. http://www.geacindia.gov.in/resource-documents/biosafety-regulations/resource-documents/Biology_of_Brinjal.pdf

[5]   Bhu, R. *Cultivation of Brinjal,* **2012**, http://agropedia.iitk.ac.in/content/cultivation-brinjal

CHAPTER 8

Analysis of Aqueous Extract of Banana Plant (*Musa balbisiana* Colla) Fiber Ash

Abstract: Materials and methods for the preparation and analysis of banana plant pseudo-stem fiber ash aqueous extract have been discussed. The extract is highly rich in potassium and carbonate, moderately in sodium and chloride. In addition, nitrate, phosphate, and 12 other heavy metals in trace concentrations have been detected. The extract is so rich in potassium and carbonate that it is a prospective renewable source for the isolation of potassium carbonate. The origin of carbonate in the extract is the oxalate present in the pseudo-stem fibers.

Keywords: Analysis of banana plant fiber ash extract, Banana plant fiber ash, Source of potassium carbonate, Use of banana plant pseudo-stem.

1. INTRODUCTION

Banana plant gives not only the delicious fruits but also provides fiber for textile as well as fiber for decorative and fancy items. All varieties of banana plants have fibers in abundance. Banana plant has long been a good source for high-quality textiles in many parts of the world, especially in Japan and Nepal. Uses of banana fibers have been discussed in chapter 1 (Section 3.3). Banana plant fibers belong to the category of Bast fiber (also called phloem fiber or skin fiber), which is the part of the phloem or bast surrounding the stem of a plant. It supports the conductive cells of the phloem and provides strength to the stem [1].

Fiber constitutes less than 5% by mass of the pseudo-stem of a matured banana plant. Crude fiber can be mechanically extracted by squeezing the pseudo-stem repeatedly to remove the fluid component. While the analysis and applications of the fluid component have been discussed in the preceding chapters, the analysis and application of the aqueous extract derived from the fiber ash are discussed in this and subsequent chapters. The fiber ash is obtained by open-air burning of the crude fibers. This work will be helpful in exploring more avenues of applications for banana plant pseudo-stems.

Dibakar Chandra Deka & Satya Ranjan Neog

The aqueous extract of the ash, which is obtained by open-air burning of the air-dried parts of a banana plant, is known as *kolakhar* in Assam, the northeastern state of India. It is widely used in Assam and other northeastern states of India by the rural folk for different purposes, including the major use as a food additive [2]. In ayurvedic literature, *kolakhar* is known as *kadalikṣâra*, which means the caustic alkali derived from the banana plant. It has an extensive application in the ayurvedic system of medicine [3].

Kolakhar, being an aqueous solution, consists of certain chemicals in water-soluble form. Many of these chemicals, being ionic in nature, are dissociated in water into their respective anions and cations. Chemical investigation shows major acid radicals present in *kolakhar* are carbonate (CO_3^{2-}) and chloride (Cl^-), and basic radicals are potassium and sodium. A couple of other basic radicals in trace concentrations are also present. It has been reported that the ash of the trunk of *Musa balbisiana* Colla contains 2.625% carbonate and 0.839% chloride in the winter sample, and the respective concentrations in the summer sample are 3.450% and 1.666%. Concentrations of potassium and sodium in the winter sample are respectively 4.650% and 0.175%, and in the summer sample, respective concentrations are 6.450% and 0.175% [4]. The aqueous extract of the banana plant fiber ash is very much like *kolakhar*, but the composition will not be exactly the same.

2. MATERIALS

The banana plant pseudo-stem consists of around 95% fluid and the residual crude fibers around 5%. Separation of the fluid and the fiber from the banana plant pseudo-stem has been discussed in Chapter 2. While the qualitative and quantitative analyses of the fluid component have been reported in Chapter 2, this Chapter will deal with the qualitative and quantitative analyses of the aqueous extracts of the ash derived from the fiber component.

A fresh piece of post-harvest banana plant trunk was taken, and the leaf sheaths and the central tender stem (floral stalk) were manually separated. The separated leaf sheaths and tender stem were washed well in running water. Juices were extracted from the leaf sheath and tender stem separately with the help of a sugarcane juicer machine by squeezing, again and again, to extract the juice as much as possible. The fibers were dried under the sun and weighed. Results are shown in Table **1**.

Table 1. Relative quantities of raw material, fluid, fiber, and ash.

Entry	Raw Mterial	Weight of Raw Mass (kg)	Volume of Juice (L)	Weight of Dry Fiber (kg)	Weight of Ash (kg)
1	Leaf sheath	25.000	17.26	0.831	0.102
2	Tender stem	9.900	5.30	0.340	0.050

After the separation of the juice, the residual fibers were the materials for study. The fibers from the leave sheath and the tender stem were separately dried under bright sunlight and burnt in the open air. The gray-colored residues left after complete burning were allowed to cool in a desiccator, weighed, and kept in airtight containers until used.

3. METHODOLOGY

3.1. Preparation of Aqueous Extract

A mixture of 25 g of dry ash and 500 mL distilled water in a 1000 mL conical flask was magnetically stirred for an hour. It was filtered, and the filtrate was made up to 500 mL. The filtrate, light yellow, looked like *kolakhar* [2] and was taken for analysis.

3.2. Measurement of pH of the Aqueous Extract

The pH of the extract was recorded with the help of a digital pH meter (Eutech Instrument, pH 510, pH/mV/°C meter). The procedure that was given in the manual was followed. The meter was calibrated using buffer capsules of pH 4, pH 7, and pH 10. The cell of the meter was immersed in the experimental solutions, and pH was recorded at ambient temperature (24°C). The aqueous extracts recorded pH in the range from 10 to 11.

3.3. Procedures Involved in the Chemical and Spectroscopic Investigation of the Aqueous Extracts

Qualitative analysis of the water extracts was performed by following the standard procedures of chemical tests [5] and confirmed the presence of the following acid and basic radicals as the major components.

Acid radicals: carbonate (CO_3^{2-}), chloride (Cl^-), nitrate (NO_3^-) and phosphate (PO_4^{3-})

Basic radicals: sodium (Na^+) and potassium (K^+)

The concentrations of potassium and sodium were estimated by flame photometry, chloride by quantitative gravimetric analysis, and nitrate and phosphate were determined by UV-Vis spectrophotometric methods. The concentrations of trace metals were determined by Atomic Absorption Spectroscopy. All these procedures are described in Chapter 2.

3.4. Quantitative Estimation of Carbonate (CO_3^{2-})

In this method, the total alkali in the given solution (carbonate + hydroxide) is determined by titration with standard acid, using methyl orange as the indicator. In the second portion of the solution, the carbonate is precipitated with a slight excess of barium chloride solution, and, without filtering, the solution is titrated with standard acid using phenolphthalein as an indicator. The latter titration gives the hydroxide content, and by subtracting this from the first titration, the volume of acid required for the carbonate is obtained [6].

50 mL of the aqueous extract derived from the fiber was taken in a 250 mL conical flask, titrated against standard 0.1 M hydrochloric acid using methyl orange as the indicator. This gives the total alkalinity (hydroxide + carbonate). Another 50 mL of the extract was taken in a conical flask and warmed to 70 °C, and 1% barium chloride solution slowly in slight excess was added until no further precipitate was observed. The solution was cooled at room temperature, followed by adding a few drops of phenolphthalein indicator and titrated very slowly with constant stirring against standard 0.1 M hydrochloric acid. As the endpoint was reached, the color of the solution became pink to colorless. The amount of acid used corresponded to the present hydroxide.

Most accurate results were obtained by considering the above titration as a preliminary one. In order to ascertain the approximate hydroxide content, another titration was carried out. Most of the hydroxide in a 50 mL portion of the extract was neutralized with standard hydrochloric acid. It was then heated and precipitated as before. Under these conditions, practically pure barium carbonate was precipitated.

$$1 \text{ mL 1M HCl} \equiv 0.0170 \text{ g OH}^-$$

$$1 \text{ mL 1 M HCl} \equiv 0.0300 \text{ g CO}_3^{2-}$$

Hence, the quantity of carbonate (CO_3^{2-}) in 500 mL of an extract derived from 25 g of ash

$$= \frac{Vol.\,of\,acid \times Normality\,of\,acid \times 30}{Vol\,of\,extract \times 2}\,g$$

%CO_3^{2-} in ash

$$= \frac{Vol.\,of\,acid \times Normality\,of\,acid \times 30 \times 2}{Vol.\,of\,extract}$$

CO_3^{2-} in extract (ppm) = %CO_3^{2-} in ash × 500

4. RESULTS AND DISCUSSION

Relative quantities of fluid, fiber, and ash from leave sheaths and tender stem are shown in Table **1**. Leaf-sheaths afforded a higher quantity of fluid and less fiber as compared to the tender stem. While the leaf sheaths afforded fluid 0.690 L/kg against 0.535 L/kg from the tender stem, the quantity of fiber from the leave sheaths was 33.24 g/kg against 34.34 g/kg from the tender stem. The ash content of fibers from leaf sheaths is lower (122.7 g/kg) as compared to that of the tender stem (147.1 g/kg).

Concentrations of major cations and anions present in the aqueous extracts from leaf sheath fiber and tender stem fiber are shown in Tables **2** and **3**, respectively. In both the extracts, K^+ and Na^+ are the two major cations. However, K^+ is 17 times of Na^+ in leaf sheath fiber ash extract against 18 times in tender stem fiber ash extract. The sodium and potassium contents in the sample from the leaf sheath fiber ash are higher than the corresponding values in the sample prepared from the tender stem fiber ash.

Among the major anions, CO_3^{2-} and NO_3^- are almost at an equal concentration in both the extracts. While Cl^- concentration is higher than PO_4^{3-} concentration in leaf sheath fiber extract, PO_4^{3-} concentration is higher than Cl^- concentration in tender stem fiber ash extract.

It is important to note that oxalate ion ($C_2O_4^{2-}$) is present but carbonate ion (CO_3^{2-}) is absent in banana plant stem juice (Chapter 2). In the aqueous extracts of fiber ashes, we have observed the presence of carbonate ion (CO_3^{2-}) but not oxalate ion ($C_2O_4^{2-}$). This implies that oxalates are structural components of fibers, and these get converted to carbonates during the burning process.

Table 2. Results of chemical and spectroscopic estimation of constituent ions in the aqueous extract of leaf sheath fiber ash.

Entry	Ion	ppm	Wt% in Ash	Methods of Determination
1	Na^+	468	0.936	Flame photometry
2	K^+	7986	15.972	Flame photometry
3	Cl^-	485	0.970	Gravimetric analysis using silver nitrate
4	CO_3^{2-}	6300	12.600	Volumetric analysis using hydrochloric acid
5	NO_3^-	26	0.052	UV- Spectrometry
6	PO_4^{3-}	105	0.210	UV-Visible Spectrometry

Table 3. Chemical and spectroscopic estimation of constituent ions in water extract of tender stem fiber ash.

Entry	Ion	ppm	Wt% in Ash	Methods of Determination
1	Na^+	428	0.856	Flame photometry
2	K^+	7856	15.712	Flame photometry
3	Cl^-	283	0.565	Gravimetric analysis using silver nitrate
4	CO_3^{2-}	6302	12.613	Volumetric analysis using hydrochloric acid
5	NO_3^-	21	0.042	UV- Spectrometry
6	PO_4^{3-}	116	0.232	UV-Visible spectrometry

As many as 12 trace metal ions have been detected and estimated in both the aqueous extracts by Atomic Absorption Spectrometry. The twelve metals that are detected and estimated are Al, Cd, Co, Cr, Cu, Fe, Mg, Mn, Ni, Pb, V, and Zn, amongst which Fe and Al levels are quite high in both the extracts as compared to other trace metals (Table **4** and **5**). It is interesting to note that the level of Al (8.643 ppm) is higher than Fe (6.536 ppm) in the aqueous extract derived from the leaf sheath fiber ash, whereas the level of Fe (9.929 ppm) is higher than that of Al (5.996 ppm) in the sample derived from the ash of the tender stem fiber.

Although there are quantitative differences, albeit low, regarding chemical constituents present in the two aqueous extracts discussed above, there is virtually no difference between the two as far the quality is concerned, and the two can be considered identical for all practical purposes, including traditional or proposed uses.

Table 4. Estimation of trace metals in the aqueous extract from the leaf sheath fiber ash of banana plant pseudo-stem (*Musa-balbisiana*) by Atomic Absorption Spectrometer.

Entry	Metal	ppm
1	Al	8.643
2	Cd	0.005
3	Co	0.043
4	Cr	0.022
5	Cu	0.083
6	Fe	6.536
7	Mg	0.582
8	Mn	0.053
9	Ni	0.801
10	Pb	0.006
11	V	0.349
12	Zn	0.532

Table 5. Estimation of trace metals in the aqueous extract of tender stem fiber ash by Atomic Absorption Spectrometer.

Entry	Metal	ppm
1	Al	5.996
2	Cd	0.008
3	Co	0.041
4	Cr	0.018
5	Cu	0.083
6	Fe	9.929
7	Mg	0.952
8	Mn	0.150
9	Ni	0.613
10	Pb	0.007
11	V	0.110
12	Zn	0.234

5. CONCLUSION

The aqueous extracts of banana plant (*Musa balbisiana*) leaf sheath fiber ash and tender stem fiber ash are potential sources of potassium carbonate and potassium chloride. Both chemicals find a huge market across the globe. There is virtually no source of potassium carbonate in nature, and natural reserves of potassium chloride are confined to a few countries only. India does not have any natural deposits of potassium chloride. Therefore, a post-harvest banana plant can be explored for the production of potassium carbonate and potassium chloride.

REFERENCES

[1]   a) Esau, K. *Anatomy of Seed Plants*; John Wiley & Sons: New York, **1977**. b) *Bast fibre, Wikipedia the free Encyclopedia,* https://en.wikipedia.org/wiki/Bast_fibre

[2]   Deka, D.C.; Talukdar, N.N. Chemical and spectroscopic investigation on *kolakhar* and evaluation of its commercial importance. *Indian J. Tradit. Knowl,* **2007**, *6*(1), 72-78.

[3]   Reddy, K.R.C. *Text Book of Rasa Sastra*; Chaukhambha Sanskrit Bhawan: Baranasi (India), **2010**.

[4]   Deka, D.C; Talukdar, N.N *Kolakhar and its Chemistry,* 2nd ed.; Techno Ed Publication: Guwahati, **2011**.

[5]   Vogel, I. *A Text Book of Macro and Semi micro Qualitative Inorganic Analysis,* 4th ed; Orient Longman: New Delhi, **1975**.

[6]   *Vogel's Textbook of Quantitative Inorganic Analysis,* 4thed.; revised by Basset, J. Denney, R.C. Jeffery, G.H. and Mendham, J.; English Language Book Society: Longman, **1986**.

CHAPTER 9

Isolation of Salt Alternative and Potassium Chloride from Banana Plant (*Musa Balbisiana* Colla) Fiber Ash

Abstract: Materials and methods for the isolation of potassium chloride from banana plant fiber ash aqueous extract have been discussed. The extract, after neutralization with hydrochloric acid, affords a solid which consists of nearly 90% of potassium chloride. The solid is a prospective table salt alternative, especially for individuals suffering from hypokalemia and high blood pressure. Fractional crystallization of the solid affords potassium chloride of 96% purity, which can find other applications.

Keywords: Banana plant fibre ash, Isolation of potassium chloride, Salt alternative in hypokalemia/hypertension, Salt substitute, Use of banana plant pseudo-stem.

1. INTRODUCTION

The chemical compound potassium chloride (KCl) is a colorless, odorless vitreous crystalline substance in its pure state. Potassium chloride crystals are face-centered cubic. It is highly soluble in water (339.7 gL^{-1} at 20 °C) and melts at 770 °C [1]. Commercially it is produced by mining from the natural deposits and labeled as 'muriate of potash'. Potash is primarily used as a fertilizer (approximately 95%) to support plant growth, increase crop yield and disease resistance, and enhance water preservation. Small quantities are used in the manufacturing of potassium-bearing chemicals such as detergents, ceramics, pharmaceuticals, water conditioners, as an alternative to de-icing salt, *etc.* Natural deposits of potash have been discussed in Chapter 1 and its importance in fertilizers have been discussed in Chapter 3. Major industrial uses of potash can be summed up as (i) 30-35% in detergents and soaps, (ii) 25-28% in glass and ceramics, (iii) 20-22% in textiles and dyes, (iv) 13-15% in the manufacture of chemicals and drugs, and (v) 7-5% in other uses [2].

Dibakar Chandra Deka & Satya Ranjan Neog

Potassium is important not only for plant health, it is equally important for human health. A low level of potassium in human blood (3.5 to 5.0 mM per liter) is responsible for a disease condition called hypokalemia. Potassium chloride is used to treat hypokalemia [3, 4]. Mild hypokalemia is often part of a poor intake of potassium in food; in such conditions, potassium-containing foods such as leafy green vegetables, tomatoes, citrus fruits, oranges, or bananas may be recommended [5]. Some cardiac surgery procedures cannot be carried out on the beating heart. For these procedures, the surgical team will bypass the heart with a heart-lung machine and inject potassium chloride into the heart muscle to stop the heartbeat.

Potassium chloride is used as a salt alternative or salt substitute for people looking to cut down their salt intake (sodium chloride), either on its own or in a mixture with sodium chloride. Sodium chloride is needed by all known animals and plants in small quantities but harmful in excess. An individual can consume not more than 1500–2300 mg of sodium (3750–5750 mg of salt) per day, depending on age [6]. Eating too much salty food causes hypertension, abnormal heart development, kidney disorder, dehydration and swelling, digestive diseases, electrolyte and hormone imbalance. Among other ways to flavor our food and limit sodium intake for lowering blood pressure and other physiological anomalies, the salt substitute is one option. Salt substitutes replace some or all of the sodium with potassium chloride in the salt. The downside of using potassium chloride as a salt substitute is that it is also harmful when consumed in large quantities. People who suffer from kidney problems can find it difficult to expel large quantities of potassium chloride from their bodies [7]. For a healthy adult, the daily recommended intake of potassium in food is about 3500 mg. Certain groups of people, like athletes and others involved in intensive physical activity, may require more.

Potassium chloride is sometimes used in water as a completion fluid in petroleum and natural gas operations. It is useful as a beta radiation source for calibration of radiation monitoring equipment because natural potassium contains 0.0118% of the isotope ^{40}K. One kilogram of KCl yields 16350 becquerels of radiation consisting of 89.28% beta and 10.72% gamma with 1.46083 MeV [8]. Along with sodium chloride and lithium chloride, potassium chloride is used as a flux for the gas welding of aluminum. It is also used in various brands of bottled water.

2. MATERIALS

The material used in the experiment was the banana plant pseudo-stem fiber ash. The required ash was prepared from the leaf sheath of banana plant pseudo-stem fiber. The fiber was isolated by the repeated mechanical squeezing of leaf sheaths with a sugarcane juicer machine. The fiber was then dried under the bright sun, followed by open-air burning. The ash was allowed to cool on its own and preserved in an airtight container until used. In the preceding chapter (Chapter 8), we have discussed the composition of the aqueous extract of leaf sheath fiber ash and the aqueous extract of tender stem fiber ash. The two extracts do not differ significantly in chemical compositions. We have chosen leaf sheath fiber ash extract only because larger quantities of leaf sheaths are easily available from post-harvest banana plants.

3. METHODOLOGY

3.1. Preparation of Aqueous Extract of Pseudo-stem Fiber Ash

A mixture of 25 g of dry ash and 500 mL distilled water in a 1000 mL conical flask was magnetically stirred for one hour and filtered. The filtrate was used for the isolation of potassium chloride (Analysis and results discussed in Sections 3 & 4, Chapter 8).

3.2. Preparation of Salt Substitute

The chemical and spectroscopic investigation on the aqueous extract of fiber ash is presented in Chapter 8 and the composition discussed. The major chemical species present in the extract are K^+, Na^+, CO_3^{2-}, Cl^- along with minor species NO^{3-} and PO_4^{3-}. The extract is strongly alkaline (pH 10.9) due to the presence of carbonates. The extract was made neutral to litmus (pH 7.0) by dropwise addition of dilute hydrochloric acid. The resulting solution was slowly evaporated to near dryness. It was then allowed to cool inside a desiccator and weighed. 500 mL of extract from 25 g ash yielded 8.5026 g of colorless and odorless solid. The chemical composition of the solid, was established by chemical and spectroscopic investigation (*cf.* Chapter 8). Results are shown in Tables **1** and **2**.

3.3. Isolation of Potassium Chloride

The solid obtained from the aqueous extract of fiber ash after neutralization with dilute HCl (*cf.* Section 3.2 above) may be used either as a salt alternative or to obtain KCl of a higher grade. KCl of higher grade (higher purity) can be derived by fractional crystallization.

Table 1. Composition of the solid isolated from the aqueous extract of banana plant pseudo-stem leaf sheath fiber ash (Amount of ash = 25g, Volume of extract = 500 mL, Total solid = 8.5026 g).

Entry	Chemical Constituents	Quantity in Gram	Quantity in %
1	Na^+	0.234	2.75
2	K^+	3.993	46.96
3	Cl-	4.012	47.19
4	NO_3^-	0.013	0.15
5	PO_4^{3-}	0.053	0.62

1000 mL aqueous extract from 50 g ash of leaf sheath fiber of banana plant (*Musa balbisiana*) was prepared and neutralized to litmus by 1 M HCl. A few more drops of HCl were added to make the solution slightly acidic *i.e.,* pH should be less than 7, to facilitate easier crystallization of KCl. The slightly acidic solution was taken in a 1000 mL beaker and slowly heated to evaporate the solvent. When the volume became 50 mL, the whole solution was transferred to another pre-weighed 100 mL beaker and heated again to reduce the volume to about 18 mL. The solution was then allowed to cool to ambient temperature. At this stage, the ionic product of K^+ and Cl⁻ ions crosses the solubility product (21.7) of KCl, but the ionic product of Na^+ and Cl⁻ cannot cross the solubility product (37.3) of NaCl. So, KCl could be precipitated at this concentration. After half an hour, a white solid was crystallized. The supernatant liquor was separated by decantation, and the white crystals were left at the bottom of the beaker. The crystals were dried by spreading over a filter paper and then keeping in a desiccator. The solid so obtained was termed as the 1ˢᵗ Crop. Its weight was 8.1142 g.

The supernatant liquor was transferred to a pre-weighed 100 mL beaker and evaporated to dryness on a hot plate. The solid obtained in the beaker was allowed to cool in a desiccator and weighed. The solid so obtained was termed as the 2ⁿᵈ Crop. Its weight was 8.8910 g.

The chemical composition of the solids from the 1st and the 2ⁿᵈ Crops were established by chemical and spectrochemical methods (*cf.* Chapter 8). The results are presented in Tables **3** to **5**.

4. RESULTS AND DISCUSSION

The aqueous extract of the ash derived from the fibers of banana plant pseudo-stem is highly alkaline with a pH measuring 10.9. The solution, after careful neutralization with dilute HCl followed by evaporation, yielding a colorless and odorless crystalline solid. The composition of the solid is shown in Table 1.

The product consists of a considerably high percentage of K^+ (46.96%) and Cl^- (47.19%) as compared to other constituents. Mass percentage of other components *i.e.* NO_3^- (0.15%) and PO_4^{3-} (0.62%), are low. Nearly 89.71% of the total mass of the solid consists of potassium chloride (KCl) and 7.00% of sodium chloride (NaCl). Other 3.29% consists of chlorides, nitrates and phosphates of trace metals such as Al, Cd, Co, Cr, Cu, Fe, Mg, Mn, Ni, Pb,V and Zn (Table 2). These trace metals are micronutrients; they play vital roles in our body functions. They act as co-factors for enzyme reactions, maintain pH within the body, maintain proper nerve function, help to contract and relax muscles, help to regulate our body's tissue growth, provide structural and functional support for the body *etc.* Moreover, little sodium and high potassium make the solid useful for human health by reducing the risk of high blood pressure and cardiovascular disease associated with the high intake of sodium chloride [9]. As the components are natural in origin, the solid obtained from the leaf sheath of banana plant (*Musa balbisiana*) pseudo-stem fiber may be the best table salt alternative.

Table 2. Trace metals in the white crystalline solid isolated from the aqueous extract of pseudo-stem fiber ash.

Entry	Metal	Mg/100g
1	Al	50.841
2	Cd	0.209
3	Co	4.253
4	Cr	0.529
5	Cu	0.488
6	Fe	38.447
7	Mg	3.424
8	Mn	0.312
9	Ni	4.712
10	Pb	0.305
11	V	2.053
12	Zn	3.129

Recommended intakes of trace metals for individuals of age group 31-70 year are: Cr 0.02 -0.035 mg/d, Cu 0.90 mg/d, Fe 8 mg/d (18 mg/d for female of age group 31-50 year), Mg 320-420 mg/d, Mn 1.8-2.3 mg/d, Zn 8-11 mg/d [10]. On the other hand, the adequate intake level of Al for an adult is 5-150 mg/d [11], whereas the provisional tolerable weekly intake (PTWI) for Cd and Pb is 0.007 mg/kg body weight and 0.025 mg/kg body weight [12] respectively. Tolerable upper intake levels of Ni and V are 1.0 mg/d and 1.8 mg/d [13] for adults. The adequate intake level for Co is 0.01-0.02 mg/d for both the male and female adults, but the safe upper limit is 1.4 mg/d. In the case of Cr, the safe upper limit is 10 mg/d [14].

Modern civilization has changed the food habits of mankind. Dependent on commercial readymade foods, intake of natural foods and vegetables by people has been considerably reduced. Natural foods and vegetables help maintain a proper balance of Na^+, K^+ and other beneficial trace metals. But commercial foods are short of this balance; they are rich in Na^+ but poor in K^+. As recommended by health experts, intake of Na^+ for a healthy adult should be limited to 1500–2300 mg per day and K^+ 4700 mg/day. The recommended daily allowance of potassium is more than double that for sodium [15], yet a typical person consumes less potassium but more sodium in a given day [16]. A person who takes relatively high dietary potassium has lower blood pressure [3]. The Third National Health and Nutrition Examination Survey (NHANES III) indicated that higher dietary potassium intakes were associated with significantly lower blood pressures [17]. In 2004, the Food and Nutrition Board of the Institute of Medicine established an adequate intake level for potassium based-on the different age groups that have been found to lower blood pressure, reduce salt sensitivity, and minimize the risk of kidney stones [15].

The white crystalline product, on careful fractionation, afforded two fractions, 1st Crop and 2nd Crop. Analyses do show that the 1st Crop consists of nearly 97% of K^+ and Cl^- while the 2nd Crop consists of about 90% of K^+ and Cl^- (Table **3**). Na^+ in the 1st Crop is only about 0.45% against 4.85% in the 2nd Crop. While the NO_3^- content was not detected in the 1st Crop, it was present in the 2nd Crop to the extent of 0.26%. The presence of PO_4^{3-} in the 1st Crop is also much lower (0.21%) as compared to that in the 2nd Crop (0.99%). Thus, the 1st Crop consists of a higher percentage of KCl and lower impurities as compared to the 2nd Crop (Table **3**).

The concentrations of trace metals in 1st and 2nd Crops, as estimated by Atomic Absorption Spectrometric method, are shown in Tables **4** and **5**. Results show that major parts of the trace metals appear in the 2nd Crop. There is 8.02mg/100g of Al in the 1st Crop against 41.28mg/100g in the 2nd Crop. Similarly, against 6.05mg/100g of Fe in the 1st Crop, there is 31.21mg/100mg in the 2nd Crop. The presence of other heavy metals, such as Cd, Co, Cr, Cu, Mn, Ni, V and Zn is also higher in the 2nd Crop than those in the 1st Crop.

Table 3. Major chemical constituents and their quantities in 1st and 2nd Crops after fractional crystallization.

Entry	Sample Code	The Total Weight (g)	Quantity of Constituents (in Gram)					Total Mass Accounted for
			Na$^+$	K$^+$	Cl$^-$	NO$_3^-$	PO$_4^{3-}$	(g)
1	1st Crop	8.1142 (47.72%)	0.0369 (0.45%)	4.0950 (50.47%)	3.8028 (46.87%)	BDL	0.017 (0.21%)	7.9517 (97.99%)
2	2nd Crop	8.8910 (52.28%)	0.4311 (4.85%)	3.8910 (43.76%)	4.1762 (46.97%)	0.023 (0.26%)	0.088 (0.99%)	8.6093 (96.83%)

Table 4. Concentration of trace metals in the 1st Crop.

Entry	Metal	mg/100g
1	Al	8.02
2	Cd	-
3	Co	0.78
4	Cr	-
5	Cu	-
6	Fe	6.05
7	Mg	0.48
8	Mn	-
9	Ni	0.60
10	Pb	0.25
11	V	0.46
12	Zn	0.39

Table 5. Concentration of trace metals in the 2nd Crop.

Entry	Metal	mg/100g
1	Al	41.28
2	Cd	0.16
3	Co	3.47
4	Cr	0.46
5	Cu	0.43
6	Fe	31.21

(Table 5) cont.....

Entry	Metal	mg/100g
7	Mg	2.83
8	Mn	0.26
9	Ni	3.96
10	Pb	-
11	V	1.55
12	Zn	2.63

5. CONCLUSION

The water extract of banana plant pseudo-stem fiber ash is a strong alkali with pH 10.9. The extract is quite rich in potassium concentration, much higher as compared to that of sodium. Chloride content is much lower than that of carbonate. After neutralization with hydrochloric acid, the resulting solution contains only chloride, no carbonate. After evaporation of the solvent from the neutralized aqueous extract, the recovered solid contains 89.71% of potassium chloride and the rest consists of Na^+, Cl^-, NO_3^-, PO_4^{3-} and trace metals Al, Fe, Cd, Co, Cu, Cr, Mg, Mn, Ni, Pb, V and Zn. The higher level of potassium chloride in the solid enhances its quality for use as a table salt alternative, especially for individuals suffering from hypokalemia and high blood pressure.

The 1st Crop recovered from the fractional crystallization of the solid (from the aqueous extract of the pseudo-stem fiber ash) contains nearly 96% of potassium chloride. This white crystalline solid can find uses in pharmaceutical and other applications requiring high purity potassium chloride.

The 2nd Crop contains nearly 83% of potassium chloride and is less pure potassium chloride as compared to the 1st Crop. Yet this fraction can possibly be used as a table salt substitute for patients having problems of hypokalemia and high blood pressure. This fraction is rich in metal micronutrients, and therefore additional benefits for individuals are expected.

REFERENCES

[1]   Potassium chloride. *Wikipedia the free encyclopedia,* https://en.wikipedia.org/wiki/Potassium_chloride

[2]   Fertilizer Manual. *UNIDO-IFDC,* **1998,** https://catalogue.nla.gov.au/Record/1864349

[3]   Barri, Y.; Wingo, C. The effects of potassium depletion and supplementation on blood pressure: A clinical review. *Am. J. Med. Sci,* **1997,** *314*(1), 37-40.

[4]   Thackaberry, E.A. Non-clinical toxicological considerations for pharmaceutical salt selection. *Expert Opin. Drug Metab,* **2012,** , 1742-5255. online

[5]   a) Krishna, G.G; Miller, E; Kapoor, S Increased blood pressure during potassium depletion in normotensive men. *N. Engl. J. Med,* **1989,** *320*(18), 1177-1182. b) Halpaerin, M.L; Kamel, K.S

Potassium. *Lancet,* **1998**, *352*, 135-140. c) Nordone, D.A; McDonald, W.J; Girard, D.E Mechanism in hypokalemia: Clinical correlation. *Medicine,* **1978**, *57*(5), 435-446.

[6]   *Dietary Guidelines for Americans,* (7th ed.), **2010**, www.dietaryguidelines.gov

[7]   Clayton, K *Uses for potassium chloride In: Healthfully,* http://www.healthyfully.com/ 200705-use--for-potassium-chloride/**2011**.

[8]   *Potassium chloride In: The catalogue of production,* http://karbohim.kz/en_1-13.html accessed 2020-07-17

[9]   *Salt and Health; Scientific Advisory Committee for Nutrition,* **2003**, www.tso.co.uk/bookshop accessed 2020-07-17

[10]   a) Trumbo, P; Yates, A.A; Schlicker, S; Poos, M Dietary reference intakes: vitamin A, vitamin K, arsenic, boron, chromium, copper, iodine, iron, manganese, molybdenum, nickel, silicon, vanadium, and zinc. *J. Am. Diet. Assoc,* **2001**, *101*(3), 294-301. b) Otsuka, Y; Isomoto, S; Noda, H Dietary intake of trace elements in the general population, estimated from a regional nutritional survey, and comparison with recommended dietary allowances and tolerable upper intake levels. *Nippon Koshu Eisei Zasshi,* **2000**, *47*(9), 809-819.

[11]   a) Kawahara, M; Kato-Negishi, M Link between aluminum and the pathogenesis of Alzheimer's disease: The integration of the aluminum and amyloid cascade hypotheses. *Int. J. Alzheimers Dis,* **2011**, 1-17. Article ID 276393. b) Klotz, K; Weistenhöfer, W; Neff, F; Hartwig, A; van Thriel, C; Drexler, H The health effects of aluminum exposure. *Dtsch. Arztebl. Int,* **2017**, *114*(39), 653-659.

[12]   *Evaluation of certain food additives and contaminants: fifty-fifth report of the Joint FAO/WHO Expert Committee on Food Additives,* https://apps.who.int/iris/handle/10665/42388 accessed 2020-07-18.

[13]   Bauer, K.D; Liou, D; Sokolik, C.A *Nutrition counseling and education skill development,* 2nd ed; Wadsworth Cengage Learning: Belmont, California, **2012**.

[14]   Leblanc, J-C; Guérin, T; Noël, L; Calamassi-Tran, G; Volatier, J-L; Verger, P Dietary exposure estimates of 18 elements from the 1st French total diet study. *Food Addit. Contam,* **2005**, *22*(7), 624-641.

[15]   Dietary reference intakes for water, potassium, sodium, chloride and sulfate. In: *Panel on Dietary Reference Intakes for Electrolytes and Water, Standing Committee on the Scientific Evaluation of Dietary Reference Intakes, Food and Nutrition Board; Institute of Medicine of the National Academies*; The National Academies Press: Washington, D.C, **2005**.

[16]   Caggiula, A.W.; Wing, R.R.; Nowalk, M.P.; Milas, N.C.; Lee, S.; Langford, H. The measurement of sodium and potassium intake. *Am. J. Clin. Nutr,* **1985**, *42*(3), 391-398.

[17]   Hajjar, I.M; Grim, C.E; George, V; Kotchen, T.A Impact of diet on blood pressure and age-related changes in blood pressure in the US population: Analysis of NHANES III. *Arch. Intern. Med,* **2001**, *161*(4), 589-593.

CHAPTER 10

Isolation of Potassium Carbonate from the Banana Plant (*Musa Balbisiana* Colla) Fiber Ash

Abstract: Materials and methods for the isolation of potassium carbonate from banana plant fiber ash aqueous extract have been discussed. The extract is highly rich in potassium and carbonate. The soluble component of the banana plant pseudo-stem fiber ash is about 32%, of which 89% is K_2CO_3. Potassium carbonate of 93% purity can be isolated from the soluble component by fractional crystallization. Thus, banana plant pseudo-stem is a viable and renewable source of potassium carbonate. There is practically no source of potassium carbonate in nature.

Keywords: Banana plant fiber ash, Isolation of potassium carbonate, Source of potassium carbonate, Use of banana plant pseudo-stem.

1. INTRODUCTION

Potassium carbonate (K_2CO_3) is a white, odorless, and crystalline inorganic salt that forms a strongly alkaline solution in water but insoluble in alcohol. It is known by many other names such as potash, carbonate of potash, pearl ash, salt of tartar, and salt of wormwood. It is deliquescent and, by absorbing moisture from the air gets easily converted to the sesquihydrate with the formula $K_2CO_3 \cdot 1.5H_2O$. There is practically no source of K_2CO_3 in nature except minor sources in a few African lakes as well as in the Dead Sea [1]. In the eighteenth century, it was first produced by leaching wood ashes into a large iron pot followed by evaporation of the resulting solution that left behind a white residue called *pot ash.* The *pot ash* eventually became known as potash. The *pot ash* or potash was a mixture of salts that included potassium carbonate and potassium chloride as major constituents. The first patent issued by the US Patent Office was awarded to Samuel Hopkins in 1790 for an improved method of making potash and pearl ash [2].

Dibakar Chandra Deka & Satya Ranjan Neog

The physical properties of potassium carbonate are

Molar mass: 138.205 g/mol

Melting point: 891 °C (1164 K)

Boiling point: decomposes

Solubility in water: 112 g/100 mL (20 °C)

A short review of the production and uses of potassium carbonate is presented in Chapter 1. Presently, potassium carbonate is commercially produced by absorption of carbon dioxide in aqueous potassium hydroxide. Potassium hydroxide, on the other hand, is produced by the electrolysis of potassium chloride. Thus, the production of potassium carbonate is an energy-intensive process. It is, therefore, desirable that renewable sources for the production of potassium carbonate be explored. In Chapter 8, we have shown that banana plant pseudo-stem fiber ash is highly rich in potassium carbonate; also, post-harvest banana plant pseudo-stem is a waste. It is, therefore considered that the isolation of potassium carbonate from banana plant pseudo-stem could be an economically viable process.

2. MATERIALS

The material used in the experiment was the banana plant pseudo-stem fiber ash. The required ash was prepared from the leaf sheath of banana plant pseudo-stem fiber. The fiber was isolated by the repeated mechanical squeezing of leaf sheaths with a sugarcane juicer machine. The fiber was then dried under the bright sun, followed by open-air burning. The ash was allowed to cool and then preserved in an airtight container until used. In Chapter 8, we have discussed the composition of the aqueous extract of leaf sheath fiber ash and the aqueous extract of tender stem fiber ash. The two extracts do not differ significantly in chemical composition. For our experimental purpose, we used leaf sheath fiber ash extract only because larger quantities of leaf sheaths were easily available from post-harvest banana plants. However, segregation or separation of the two is not generally required.

3. METHODOLOGY

The detection and estimation of the chemical constituents in the aqueous extract of the leaf sheath fiber ash are discussed in Chapter 8. The two major constituents present are K^+ and CO_3^{2-}.

3.1. Preparation of Aqueous Extract of Pseudo-stem Fiber Ash

A mixture of 25 g of dry ash and 500 mL distilled water in a 1000 mL conical flask was magnetically stirred for an hour and filtered. The filtrate was used for the isolation of potassium chloride (*cf.* Section 3.1, Chapter 8).

3.2. Composition of the Solid from the Aqueous Extract of the Fiber Ash

500 mL of an aqueous extract derived from 25 g of banana plant pseudo-stem fiber was slowly evaporated in a 1000 mL beaker over an electric hot plate. When the volume of the solution became about 60 mL, it was transferred into a pre-weighed 100 mL beaker. Again the solution was evaporated to dryness to yield a white solid. The solid so obtained was kept in a furnace at 250 to 350 °C for dehydration and weighed. The composition of the solid was estimated based on the concentration of the ions in the aqueous extract (*cf.* Chapter 8), and results are shown in Tables **1** and **2**.

3.3. Isolation of High-Grade Potassium Carbonate from Aqueous Extract of Fiber Ash

From the study of the solubility product of salts [3], it is noted that the solubility product of sodium carbonate (K_2CO_3) is 2130, which is very high as compared to that of sodium chloride (37.3), potassium chloride (21.7), and sodium carbonate (1.2) at 25°C. The concentration of major constituents in the aqueous extract derived from the fiber ash is as follows (Chapter 8):

$[Na^+] = 0.0203$ mol/L; $[K^+] = 0.2042$ mol /L

$[Cl^-] = 0.0137$ mol/L; $[CO_3^{2-}] = 0.1050$ mol/L

$[NO_3^-] = 0.0004$ mol/L; $[PO_4^{3-}] = 0.0011$ mol/L

From the relationship between the solubility product and ionic product of salts, it can be predicted that if the volume of the extract is reduced to 2.5% of its original volume, the ionic product will exceed the solubility product of sodium carbonate (Na_2CO_3) only; as a result Na_2CO_3 will be precipitated. But potassium carbonate (K_2CO_3) has a very high solubility product, so the ionic product cannot cross its solubility product at this concentration. The concentration of chloride ion (Cl^-) is not large enough to make the ionic products of sodium chloride and potassium chloride to exceed the corresponding solubility products. Therefore, potassium carbonate will remain in the solution and along with Cl^- ions and a part of Na^+

ions. Moreover, at this stage, almost all trace metals will precipitate either as phosphates or as carbonates because their solubility products as carbonate or phosphate are very low. With this principle in mind, the separation of potassium carbonate from its solution by fractional crystallization was planned.

3.4. The Procedure for the Isolation of K_2CO_3

1000 mL of aqueous extract prepared from 50 g of banana plant (*Musa balbisiana*) pseudo-stem fiber ash was slowly evaporated in a 1000 mL beaker by heating over an electric hot plate. When the volume of the solution was reduced to about 80 mL, the solution was transferred into a pre-weighed clean 100 mL beaker. The solution was again slowly evaporated, and the volume was reduced to 20 mL (2.0% of its original volume). When the solution was allowed to cool down to room temperature, a white solid appeared. The supernatant liquor was separated carefully with a dropper and the solid residue was dried over a low temperature electric hot plate, cooled down to room temperature in a desiccator, and weighed. The solid so obtained was termed as the 1st Crop.

The supernatant liquor was transferred to a pre-weighed 100 mL beaker and evaporated to near dryness over a low temperature electric hot plate. The solid mass in the beaker was dried in a hot air oven at 200 °C for an hour, allowed to cool down to room temperature in a desiccator, and weighed. The solid mass so obtained was termed as the 2nd Crop.

The compositions of the solids in the 1st Crop (0.9178 g) and the 2nd Crop (14.9318g) were estimated by chemical and spectrometric methods as described in Chapter 2 and Chapter 8, and the results are given in Tables 3 and 4.

4. RESULTS AND DISCUSSION

The weight of the anhydrous solid isolated from the aqueous extract of 25 g of fiber ash was 7.9248 g. The major constituents in the solid are shown in Tables 1 and 2. It is observed that 25 g of banana pseudo-stem fiber ash gives 7.9248 g of solid mass, which is 31.70% of the ash, ie, almost one-third of the total mass of the ash is water-soluble. The solid contains the highest percentage of potassium as the cation (50.39%) and the highest percentage of carbonate (CO_3^{2-}) as an anion (39.75%). The remaining part consists of Na^+ (2.95%), Cl^- (3.07%), NO_3^- (0.16%), PO_4^{3-} (0.67%) and trace metals.

Table 1. Composition of the solid isolated from the aqueous extract of banana plant pseudo-stem fiber ash.

Entry	Chemical Constituents	Quantity in Gram	Quantity in %
1	Na^+	0.234	2.95
2	K^+	3.993	50.39
3	Cl^-	0.243	3.07
4	CO_3^{2-}	3.150	39.75
5	NO_3^-	0.013	0.16
6	PO_4^{3-}	0.053	0.67

Weight of ash = 25 g; Vol. of extract = 500 mL; Solid isolated = 7.9248 g

Table 2. Trace metals in the solid isolated from the aqueous extract of leaf sheath fiber ash.

Entry	Trace Metal	Quantity (mg/100g)
1	Al	54.531
2	Cd	0.032
3	Co	0.271
4	Cr	0.139
5	Cu	0.524
6	Fe	41.237
7	Mg	3.672
8	Mn	0.334
9	Ni	5.054
10	Pb	0.038
11	V	2.202
12	Zn	3.357

The experimental findings reveal that 25 g of ash derived from a banana plant (*Musa balbisiana*) pseudo-stem can give as much as 7.057 g of K_2CO_3 (28% of ash). This means that 7.057 g K_2CO_3 can be obtained from 203.68 g dry fiber derived from 6.1 kg banana plant leaf sheaths. Hence, 3.46% potassium carbonate by weight can be obtained from leaf sheath dry fiber.

Results of fractional crystallization to isolate potassium carbonate in more pure form from the solid isolated from the aqueous extract of leaf sheath fiber ash are shown in Tables **3** and **4**.

Analysis of the results shown in Table **3** shows that out of the total soluble mass of 15.8496 g present in 50 g of ash (1000 mL aqueous extract), 0.9178 g (5.79%)

is recovered in the 1st Crop and the rest 14.9318 g (94.21%) in the 2nd Crop. Major parts of K^+ (98.36%) and CO_3^{2-} (96.81%) are found in the 2nd Crop, whereas major parts of Na^+ (56.99%) and PO_4^{3-} (67.89%) are found in the 1st Crop. Chloride ion is less in the 1st Crop (45.22%) and higher in the 2nd Crop (54.78%). Table 4 shows the relative percentage of the constituent ions in the two fractions. K^+ and CO_3^{2-} together account for 93.46% of the total mass of the 2nd Crop against 36.15% in the 1st Crop. Thus, the process described in this section for isolation of K_2CO_3 is a viable process.

Table 3. Composition of 1st and 2nd Crops from the fractional crystallization of the solid isolated from the aqueous extract of leaf sheath fiber ash.

Entry	Sample code	Total weight (g)	Quantity (in g)						Total mass accounted for	
			Na^+	K^+	Cl^-	CO_3^{2-}	NO_3^-	PO_4^{3-}	g	%
1	1st Crop	0.9178	0.2665	0.1306	0.2190	0.2012	BDL	0.0702	0.8875	96.70
2	2nd Crop	14.9318	0.2011	7.8550	0.2653	6.1000	0.0250	0.0332	14.4796	96.97

Total solid = 15.8496 g; 1st Crop = 0.9178 g; 2nd Crop = 14.9318 g

Table 4. Relative Composition of 1st and 2nd Crops from the fractional crystallization of the solid isolated from the aqueous extract of leaf sheath fiber ash.

Entry	Chemical Constituent	Quantities in the 1st Crop (%)	Quantities in the 2nd Crop (%)
1	Na^+	56.99	43.01
2	K^+	1.64	98.36
3	Cl^-	45.22	54.78
4	CO_3^{2-}	3.19	96.81
5	NO_3^-	BDL	100
6	PO_4^{3-}	67.89	32.11

5. EFFECT OF TEMPERATURE ON THE CARBONATE CONTENT OF THE ASH

It has been observed that banana plant pseudo-stem juice has no carbonate content, but it has a considerable amount of oxalate (*cf.* Chapter 2). On the other hand, the water extract of pseudo-stem fiber ash contains a high amount of carbonate but no oxalate. The presence of the high amount of carbonate and the absence of oxalate in the aqueous extract of fiber ash implies that oxalate is converted to carbonate during the process of burning. Therefore, an attempt has been made to know the effect of temperature on the conversion of oxalate to carbonate.

Oxalates are common constituents of plants and are found in the forms of oxalic acid, soluble salts of potassium, sodium, and magnesium, and also in the form of insoluble salt of calcium [4]. The amount of oxalate in plants ranges from a few percent up to 80% of the total weight of the plant. Soluble oxalates usually accumulate within the vacuoles of plant cells, and the insoluble oxalates in the cell walls of some plants. Since plant cells generally have a large vacuolar compartment, often 75 to 90% of the cell volume, massive oxalate accumulation is possible in these vacuolar compartments [4].

It is reported that oxalic acid and oxalates were detected in varying quantities in all parts of most plants, leaves, leaf stalks, flowers, tubers and roots [5]. The oxalate content of plants may also vary according to their age, the season, the climate, and the type of soil. In some plants like bananas, sugar beet leaves, spinach, a large increase in oxalate content is observed during the early stages of development, but in some other plants like rhubarb, oxalate content tends to increase as the plants mature.

To study the temperature dependency of carbonate content in the aqueous extract of the banana plant pseudo-stem fiber ash, the extracted dry fiber was burnt at different temperatures *viz.* 300, 400, 500 and 600 °C in a muffle furnace as well as in open air.

5.1. Procedure

For burning the fiber, a silica crucible of appropriate size was first cleaned and dried for an hour in a hot air oven at 150 to 200 °C, allowed to cool down to ambient temperature in a desiccator and weighed. The process of heating and cooling was repeated until a constant weight of the crucible was recorded. The crucible with 6.00 g of dry fiber was then put inside a muffle furnace and the temperature raised up to the desired level at the rate of $10°C$/min. After an hour, the crucible was taken out of the furnace, allowed to cool down to ambient temperature inside a desiccator and weighed. The ash so obtained was taken in a 250 mL conical flask, 100 mL distilled water added and stirred magnetically for 30 minutes. The mixture was filtered and the filtrate was made up to 100 mL in a volumetric flask. The carbonate content of the extract was estimated by the procedure as described in Chapter 2 and Chapter 8.

Table 5. Effect of temperature on carbonate content of banana plant.

Entry	Wt. of Dry Fiber (g)	Temp. (°C)	Wt. of Ash (g)	Color of Extract	Wt. of CO_3^{2-} (g)	CO_3^{2-} in Ash (%)
1	6.00	300	2.0944	Deep brown	-	-

(Table 5) cont.....

Entry	Wt. of Dry Fiber (g)	Temp. (°C)	Wt. of Ash (g)	Color of Extract	Wt. of CO_3^{2-} (g)	CO_3^{2-} in Ash (%)
2	6.00	400	0.8703	Brown	0.0725	9.33
3	6.00	500	0.6545	Colourless	0.0595	9.09
4	6.00	600	0.6374	Colourless	0.0567	8.89
5	6.00	Burned in open air	0.7341	Light brown	0.0925	12.60

Results in Table **5** indicate that the quantity of carbonate in the ash decreases with increasing the temperature at which the fiber is burnt. The highest quantity of carbonate was obtained when the fiber was burnt in the open air. In the case of burning at 300 °C, the color of the extract was deep brown and therefore, it was not possible to estimate the carbonate content by the titrimetric method using methyl orange indicator. The deep brown color of the extract at 300 °C indicates the presence of carbon particles to a considerable extent, and the high weight of ash indicates the incomplete combustion of fiber. The change of color of the extract from brown (at 400 °C) to light brown (in the open air) to colorless (at 500 °C) indicates the removal of carbon from the ash at higher temperatures. The weight loss of ash at higher temperatures is due to the removal of carbon particles as well as due to the thermal decomposition of metal oxalates into corresponding metal carbonates. As a result, a large increase of carbonate content in the ash and the corresponding aqueous extract is observed. The decomposition temperature of $Na_2C_2O_4$ is above 290 °C and that of $K_2C_2O_4$ is 347 °C [6]. Na_2CO_3 starts decomposition at 400 °C, but K_2CO_3 remains stable up to its melting temperature of about 899 °C. The weight loss of ash at temperature 500 °C is due to the removal of carbon and thermal decomposition of calcium oxalate (CaC_2O_4) into calcium carbonate ($CaCO_3$). Decomposition of CaC_2O_4 starts at 374 °C [7]. At temperature 500 °C and above, weight loss of ash and a decrease in carbonate content is due to the decomposition of some carbonates. The insolubility of calcium carbonate in water and decomposition of iron carbonate and aluminum carbonate into respective oxide are also contributing factors towards the decrease of carbonate in the ash at 500 °C. Decomposition of iron carbonate starts at 280 °C [8]. At 600 °C, the extract is colorless due to the absence of carbon particles, and the weight loss arises from the decomposition of unstable carbonates and the removal of volatile compounds. Calcium is not observed in the extracts because, in the presence of large excess of carbonate, calcium is completely precipitated as its solubility in water is very low (0.00003932 g/100g).

$$Na_2C_2O_4 \xrightarrow{\text{250°C–270°C}} Na_2CO_3 \xrightarrow{\text{>400°C}} Na_2O$$

$$K_2C_2O_4 \xrightarrow{\text{346.9°C–420°C}} K_2CO_3 \xrightarrow{\text{850°C}} K_2O$$

$$CaC_2O_4 \xrightarrow{\text{450°C–500°C}} CaCO_3 \xrightarrow{\text{635°C–865°C}} CaO$$

The highest carbonate continent is observed in the extract from the ash of open-air burning of the fiber. The light yellow color of the extract is due to the presence of carbon particles and ferric ions. The decomposition of oxalate compounds is strongly affected by atmospheric conditions. The supply of oxygen gas or the ease of carrying off the resulting gases plays an important role in the decomposition mechanism [8]. This may be one of the reasons of getting a high amount of carbonate in open-air burning of fiber. Therefore, the burning of banana plant pseudo-stem fiber in the open air is the best method to produce the optimum amount of carbonate.

6. CONCLUSION

Potassium carbonate is an important commercial commodity for use not only in industry but also in the pharmaceutical and agricultural sectors. There is practically no source of potassium carbonate in nature except minor sources in African lakes and the Dead Sea. It is industrially produced from potassium chloride, and the process is an energy-intensive one.

The banana plant is an excellent renewable natural source of potassium carbonate. A considerable amount of potassium carbonate can be isolated from banana plant pseudo-stem or its fiber ash. Burning of fiber in the open air is the most effective process from where the maximum quantity of carbonate can be isolated. The source of origin for carbonate might be the oxalate content in the plant's cells.

The soluble component of the banana plant pseudo-stem fiber ash is 31.70%, of which 89.15% is K_2CO_3. Potassium carbonate of 93.07% purity can be isolated from the soluble component by fractional crystallization. Thus, banana plant pseudo-stem is a viable and renewable source of potassium carbonate.

REFERENCES

[1]   *Ullman's Encyclopedia of Industrial Chemistry,* 6th ed; Wiley-VCH: Germany, **2003**, Vol. 29, pp. 93-160.

[2] a) Holt, K. *Pearl Ash, Potash, and The Ashery,* **2018**. www.KristinHolt.com. b) Hopkins, S. Improvement in the making of pot ash and pearl ash. US Patent X000001, 1790.

[3] *Solubility Products of Selected Compounds,* **2017**, http://www.saltlakemetals.com/SolubilityProducts/

[4] Franceschi, V.R.; Loewus, F.A. Oxalate biosynthesis and function in plants and fungi. In: *Calcium Oxalate in Biological Systems,* 1ˢᵗ ed.; Khan, S. R., Ed.; CRC Press, **1995**.

[5] a) Srivastava, S.K.; Krishnan, P.S. Oxalate content of plant tissues. *J. Sci. Industr.,* **1959**, *18C*, 146-148. b) Libert, B.; Franceschi, V.R. Oxalate in crop plants. *J. Agric. Food Chem.,* **1987**, *35*, 926-938.
[http://dx.doi.org/10.1021/jf00078a019]

[6] Higashiyama, T.; Hasegawa, S. The differential thermal analysis of potassium oxalate. *Bull. Chem. Soc. Jpn.,* **1971**, *44*, 1727-1730.
[http://dx.doi.org/10.1246/bcsj.44.1727]

[7] Hourlier, D. Thermal decomposition of calcium oxalate: beyond appearances. *J. Therm. Anal. Calorim.,* **2018**.
[http://dx.doi.org/10.1007/s10973-018-7888-1]

[8] Brearley, A.J. Magnetite in ALH 84001: An origin by shock-induced thermal decomposition of iron carbonate. *Meteorit. Planet. Sci.,* **2003**, *38*(6), 849-870.
[http://dx.doi.org/10.1111/j.1945-5100.2003.tb00283.x]

CHAPTER 11

Scope for Further Research and Development

Banana plant gives fruits only once, and after harvesting it leaves behind waste biomass which is 5 to 10 times the weight of the fruits. Bananas account for about 18% of the global edible fruits production, and annual global banana production amounts to about 155 MMT as per an estimate for the year 2017-18 [1]. Thus, banana farming generates every year huge quantity of biomass which finds no use rather poses a formidable disposal problem to farmers. In the preceding chapters, we have discussed how to use this huge volume of biomass for the economic benefits of farmers. In other words, 'waste to wealth' and value-addition to banana farming have been explored. However, our discussion in the preceding chapters is limited only to one aspect of agricultural use that too with only five varieties of crops. Scope for further research and development on more applications is still wide open. Hints to some of the prospective applications and related research avenues are discussed in this chapter.

1. AGRICULTURAL USE OF BANANA PLANT PSEUDO-STEM JUICE ACROSS ALL CROPS AND SOIL TYPES

In the preceding chapters, we have discussed the use of banana plant pseudo-stem juice as the alternative of muriate of potash (MOP) in the cultivation of five different crops. It is observed that banana plant pseudo-stem juice is a better alternative to MOP for the improvement of yields, but improvements in crop yields vary from one crop to another. For example, 10% improvement of yield is observed in rice cultivation against 60% improvement in the cultivation of wheat. It is therefore imperative that we examine the use of banana plant pseudo-stem juice in the cultivation of some other crops, especially in different soil types. Further, the effect of banana plant pseudo-stem juice on soil ecosystem may also be examined and compared with that of chemical MOP. This would enhance farmers' confidence and planning for the use of banana plant pseudo-stem juice in agriculture across different crop varieties in different soil ecosystems.

A couple of trace metals are present in banana plant pseudo-stem juice. Some of these trace metals are known micronutrients for plants [2 - 4]. Importance of the trace metals in plant health and productivity needs to be examined. The presence of plant micronutrients in soil varies for different soil types. Trace metals present in banana plant pseudo-stem juice may supplement essential micronutrients, especially in agricultural soils deficient in these trace metals.

2. BANANA PLANT PSEUDO-STEM IN FOOD AND MEDICINE

Banana plant pseudo-stem is known for its use as food as well as in traditional medicine. For example, the tender inner stem (also called the pith, Fig. (**1.4**) in Chapter 1) of a matured banana plant is used as food because of its health benefits and medicinal values. Eating banana stem in regular intervals is believed beneficial for those who are on a weight loss program [5]. The liquid extract of banana plant pseudo-stem is believed to be a remedy for urinary and digestive disorders [6]. It is also believed to keep high blood pressure and diabetes under control, maintain fluid balance within the body and normalize stomach upset. It is a diuretic and helps to detoxify the body [7]. Its extract is believed useful in dissolving kidney stone and stone in urinary bladder. It has been found to suppress the formation of oxalate-associated kidney stones in animal experiments, and may be a useful agent in the treatment of patients with hyperoxaluric calcium urolithiasis. Houghton and coworkers have reported that stem juice of banana plant has antivenom action [8, 9]. It is believed that mixing sap of the banana plant pseudo-stem with coconut water makes an effective treatment of nervous insomnia, epilepsy, dysentery, vomiting and hysteria in traditional Indian medicine [10]. All these and many other reported applications are yet to be scientifically validated. Wide range of scope therefore exists to look for the origin and biochemistry of reported medicinal and food values.

Apart from inorganic ions, the presence of several organic molecules in the banana plant pseudo-stem juice is indicated in our experiments (Chapter 2). Two organic molecules, which are hitherto known as synthetic origin, have been isolated and characterized. Banana plant pseudo-stem juice appears to be the first reported natural source for these two molecules. Bioactivity study of these two molecules and relating them with the medicinal use of banana plant pseudo-stem juice may unfold some other mysteries. Isolation, identification and characterization of other organic molecules present in the juice along with their bioactivity study are also now open for research. Connectivity of the identified and unidentified molecules with traditional applications may open up wider applications for banana plants.

3. SALT SUBSTITUTE FOR HYPOKALEMIA AND RELATED DISEASES

Potassium is important not only for plant health, but it is also equally important for human health. Low level of potassium in human blood is responsible for a disease condition called hypokalemia. Potassium chloride is used to treat hypokalemia [11, 12]. Potassium chloride in a mixture with sodium chloride can be used as a salt alternative or salt substitute for people looking to cut down their salt intake (sodium chloride). Salt substitute replaces some or all of the sodium with potassium in the salt. Among other ways to flavor our food and limit sodium intake for lowering blood pressure and other physiological anomalies, salt substitute is one option. Some cardiac surgery procedures cannot be carried out on the beating heart. For these procedures, the surgical team will bypass the heart with a heart-lung machine and inject potassium chloride into the heart muscle to stop the heartbeat. In Chapter 9, isolation of salt substitute and potassium chloride from banana plant pseudo-stem fiber ash has been discussed. However, the methodology discussed is of academic interest only; it is unlikely to be economically viable. Further research in this front to develop economically viable alternative procedure would be helpful.

4. ISOLATION OF POTASSIUM CARBONATE

No source of K_2CO_3 in nature is available except minor sources in a few African lakes as well as in Dead Sea [13]. In ancient Europe, it was produced by leaching wood ashes. Presently potassium carbonate is commercially produced by absorption of carbon dioxide in aqueous potassium hydroxide. Potassium hydroxide, on the other hand, is produced by the electrolysis of potassium chloride. Thus, the production of potassium carbonate is an energy intensive process. It is, therefore desirable that renewable sources for the production of potassium carbonate be explored. In Chapter 10, we have discussed the isolation of potassium carbonate from banana plant pseudo-stem fiber ash. However, the methodology is highly energy intensive and unlikely to be economically viable for commercial production. It is therefore important that further research and development in this front is necessary to find out economically viable alternative procedure for the isolation of potassium carbonate from banana plant pseudo-stem fiber ash. Banana plant pseudo-stem fiber ash extract is highly rich in potassium carbonate and therefore, can be a substitute for potassium carbonate in some uses. Potassium carbonate has plenty of uses in confectioneries, pharmaceutical industries, R & D laboratories, industries, *etc.*

5. PROSPECTIVE RESEARCH AVENUES TO FIND MORE APPLICATIONS

Banana plant pseudo-stem fiber ash is largely similar to '*kolakhar*' ash. '*Kolakhar*' is a traditional food additive [14] and is known to help in normalizing digestive disorder of stomach. In Ayurvedic literature, extensive use of *kolakhar* is described for the use against different disease conditions [15]. In ancient rural Assam and also in other North-Eastern states of India, *kolakhar* was widely and extensively used as soaps and detergents for washing cloths and shampooing hairs. It is reported that washing and cleansing with *kolakhar* help to grow and maintain long and healthy hair Fig. (**1.5**) in Chaper 1) [16]. Many other uses of *kolakhar* are known in the rural northeast [17]. In addition to the traditional uses, many modern day uses of *kolakhar* are possible [17]. A good number of modern day uses having potential for large scale commercial exploitation of *kolakhar* have been recently reported. *Kolakhar* ash has been tested in laboratory as an excellent heterogeneous catalyst in the production of biodiesel from oils and fats [18 - 20]; it appears a potential catalyst for future biodiesel industries. Uses of *kolakhar* as catalysts or basic aqueous media to accomplish useful organic transformations have also been reported [21 - 24]. Banana plant fiber ash aqueous extract is similar to *kolakhar* ash aqueous extract. Their compositions are also similar but certainly, they differ in the concentrations of constituents (Chapter 8). Therefore, the banana plant pseudo-stem fiber ash and its aqueous extract can be explored for uses similar to those of *kolakhar* ash and its aqueous extract. All these prospective applications are subject to scientific validation. Therefore, further research in these aspects would be highly appreciated.

REFERENCES

[1]   FAO © Statista. **2020**. accessed 2020-02-11.

[2]   Rout, G.R.; Sahoo, S. Role of iron in plant growth and metabolism. *Rev. Agric. Sci.,* **2015**, *3*, 1-24.
       [http://dx.doi.org/10.7831/ras.3.1]

[3]   a) White, P.J.; Broadley, M.R. Calcium in plants. *Ann. Bot.,* **2003**, *92*(4), 487-511.
       [http://dx.doi.org/10.1093/aob/mcg164] [PMID: 12933363]. b) The importance of calcium.TETRA
       Technologies. **2005-2018**. www.tetrachemicals.com/Products/Agriculture/The_Importance_of_Calci
       um

[4]   a) Hauer-Jákli, M.; Tränkner, M. Critical leaf magnesium thresholds and the impact of magnesium on
       plant growth and photo-oxidative defense: A systematic review and meta-analysis from 70 years of
       research. *Front. Plant Sci.,* **2019**, *10*, 766.
       [http://dx.doi.org/10.3389/fpls.2019.00766] [PMID: 31275333]. b) Guo, W.; Chen, S.; Hussain,N.;
       Cong, Y.; Liang, Z.; Chen, K. Magnesium stress signaling in plant: just a beginning. *Plant Signal.
       Behav.,* **2015**, *10*(3), e992287.
       [http://dx.doi.org/10.4161/15592324.2014.992287] [PMID: 25806908]

[5]   The encyclopedia of crafts in Asia Pacific Region (APR): traditional handmade products.
       https://encyclocraftsapr.com/

[6]   Lakshmana, K.; Swaminathan, S. *Speaking of Nature Cure*; Sterling Publishers Pvt. Ltd: Greater
       Noida, **2011**.

[7]   Prasad, K.V.S.R.G.; Bharathi, K.; Srinivasan, K.K. Evaluation of Musa (Paradisiaca Linn. Cultivar) –
       'Puttabale' stem juice for antilithiatic activity in albino rats. *Indian J. Physiol. Phannacol.,* **1993**,
       37(4), 337-341.

[8]   Houghton, P.J.; Osibogun, I.M. Flowering plants used against snakebite. *J. Ethnopharmacol.,* **1993**,

39(1), 1-29.
[http://dx.doi.org/10.1016/0378-8741(93)90047-9] [PMID: 8331959]

[9]   Houghton, P.J.; Skari, K. The effect of Indian plants used against snakebite on blood clotting. *J. Pharm. Pharmacol.,* **1992**, *44*, 1054-1060.

[10]   Panda, H. *HerbAl Foods and its Medicinal Values*; National Institute of Industrial Research: Delhi, **2003**.

[11]   Barri, Y.M.; Wingo, C.S. The effects of potassium depletion and supplementation on blood pressure: a clinical review. *Am. J. Med. Sci.,* **1997**, *314*(1), 37-40.
[PMID: 9216439]

[12]   Thackaberry, E.A. Non-clinical toxicological considerations for pharmaceutical salt selection. In: *Expert Opin. Drug Metab. Toxicol*; Informa UK Ltd, **2012**.
[http://dx.doi.org/10.1517/17425255.2012.717614]

[13]   *Ullman's Encyclopedia of Industrial Chemistry,* 6th ed; Wiley-VCH: Germany, **2003**, Vol. 29, pp. 93-160.

[14]   Deka, D.C.; Talukdar, N.N. Chemical and spectroscopic investigation of *kolakhar* and its commercial importance. *Indian J. Tradit. Knowl.,* **2007**, *6*(1), 72-78.

[15]   Reddy, K.R.C. *Bhaisajya KalpanāVijñānam (A Science of Indian Pharmacy),* 2nd ed.; Chaukhambha Orientalia, Sanskrit Bhawan: Varanasi, **2001**, p. 350.

[16]   Staff report on use of kolakhar. *Āmār Asom (An Assamese Daily),* **2002**.

[17]   Direct information from people on traditional knowledge practiced in rural assam. **2001**.

[18]   Deka, D.C.; Basumatary, S. High quality biodiesel from yellow oleander (*Thevetia peruviana*) seed oil. *Biomass Bioenergy,* **2011**, *35*, 1797-1803.
[http://dx.doi.org/10.1016/j.biombioe.2011.01.007]

[19]   Basumary, S.; Deka, D.C.; Deka, D.C. Composition of biodiesel from *Gmelina arborea* seed oil. *Adv. Appl. Sci. Res.,* **2012**, *3*(5), 2745-2753.

[20]   Basumatary, S.; Barua, P.; Deka, D.C. *Gmelina arborea* and *Tabernaemontana divaricata* seed oils as non-edible feedstocks for biodiesel production. *Int. J. Chemtech. Res.,* **2014**, *6*(2), 1440-1445.

[21]   Dwivedi, K.D.; Borah, B.; Chowhan, L.R. Ligand free one-pot synthesis of pyrano[2,3-*c*]pyrazoles in water extract of banana peel (WEB): A green chemistry approach. *Front Chem.,* **2020**, *7*, 944.
[http://dx.doi.org/10.3389/fchem.2019.00944] [PMID: 32039156]

[22]   Sarmah, M.; Mondal, M.; Bora, U. Agro-waste extract based solvents: Emergence of novel green solvent for the design of sustainable processes in catalysis and organic chemistry. *ChemistrySelect,* **2017**, *2*, 5180-5188.
[http://dx.doi.org/10.1002/slct.201700580]

[23]   Baruah, P.R.; Ali, A.A.; Saikia, B.; Sarma, D. A protocol for ligand free Suzuki–Miyaura cross-coupling reactions in WEB at room temperature. *Green Chem.,* **2015**, *17*(3), 1442-1445.
[http://dx.doi.org/10.1039/C4GC02522A]

[24]   Konwar, M.; Ali, A.A.; Sarma, D. A green protocol for peptide bond formation in WEB. *Tetrahedron Lett.,* **2016**, *57*(21), 2283-2285.
[http://dx.doi.org/10.1016/j.tetlet.2016.04.041]

Glossary

AgCl: Silver chloride. An important compound of metal silver. Insoluble in water.

AgNO$_3$: Silver nitrate. An important laboratory reagent. Soluble in water.

Antioxidants: Antioxidants are substances capable of destroying harmful free-radicals in our body. Free-radicals are believed to be responsible for ageing processes and many health issues. Antioxidants are considered important nutraceuticals.

Atomic Absorption Spectroscopy (AAS): AAS is a widely used analytical technique. It is used for the estimation of metallic ions in a sample. The technique is based on the principle of absorption of radiation frequency which is characteristic of atoms or ions.

Banana 'hanging cluster' 'hands' and 'fingers': A single banana fruit is referred to as 'banana finger'. A few fingers are clubbed together to make a bunch of fruits which is referred to as a 'banana hand'. Several 'banana hands' are arranged around a 'floral rachis' to build a 'hanging cluster' (see Fig. **1.3**).

Banana inflorescence: It is also referred to as banana flower spike or 'banana heart'. It houses all male and female flowers under the cover of flower bracts (see Fig. **1.3**).

Banana plant juice (BPJ): The banana plant pseudo-stem is mechanically hard-pressed to extrude the sap. The juice is highly rich in potassium.

Banana pseudo-stem (BPS): Literally pseudo-stem means false-stem. An example is banana plant. In scientific definition, the trunk of a banana plant is not a stem. But for common people the trunk looks and functions like a stem. Therefore, the trunk of a banana plant is a false-stem or pseudo-stem. The word pseudo stands for false or pretentious.

***Bhimkol* or *Athiakol*:** The banana variety with a lot of seeds in its fruits is called *bhimkol* or *athiakol* in Assamese language. Scientific name is *Musa balbisiana* Colla. The *kol* means the banana fruit. The adjective *bhim* is prefixed to mean the big and stout size of the fruit (see Fig. **1.1**).

Biodiesel: A synthetic alternative for diesel fuel derived from petroleum. Biodiesel is produced from fats and oils.

CaC$_2$O$_4$.H$_2$O: Chemical representation for calcium oxalate monohydrate. Calcium oxalate is a substance which takes up one water molecule during crystallization. It is an important phyto-chemical.

CaCO$_3$: Chemical representation for calcium carbonate. A substance insoluble in water.

CDCl$_3$: Deuterated chloroform. It is a widely used solvent in NMR spectroscopy.

CE: Catechin Equivalent. Catechin is an important phenolic compound found in plant. Plant based phenolic compounds are good antioxidants and catechin is often used as the reference compound to judge antioxidant activity of food items and nutraceuticals.

^{13}C NMR: One type of diagnostic technique that helps structural elucidation of chemical molecules by detecting the presence of ^{13}C-isotopes in the molecule.

Cooking bananas: The variety of banana which is suitable for eating after cooking. This variety of banana is also referred to as plantain. The other variety of banana which is allowed to ripe and then eaten is called sweet banana or dessert banana also called Cavendish banana. There are certain varieties of bananas which are used both as cooking banana as well as dessert banana.

Corm: It is the morphological part of a banana plant between the roots and the trunk (pseudo-stem). In true sense corm is the stem and remains close to ground. Banana corm is also banana rhizome.

FAOSTAT: It refers to the statistical data provided by the Food and Agricultural Organization of United States of America.

Flame Photometry: It is an analytical technique commonly used for the estimation of alkali metals in samples. It takes help of a flame to vaporize the ions present in samples.

Flavonoids: Flavonoids are a sub-group of phenolics. These are important phyto-chemicals.

Floral stalk or peduncle: Part of the stem which is visible between the trunk and the hanging cluster of bananas (see Fig. **1.3**).

Foliage: Petiole (Midrib) and leaf blade together make the foliage. In brief foliage makes the whole banana leave.

FT-IR spectrometer: It is a diagnostic tool used for the study of molecular structural characteristics. The tool takes help of absorption of infra-red radiation by molecules.

GAE: Gallic Acid Equivalent. Gallic acid is an important phyto-phenol and excellent antioxidant. Often used as the reference substance to judge the antioxidant property of foods and other nutraceuticals.

GC: Gas Chromatography. It is an important chromatographic technique which can be used to identify volatile organic molecules with high precision.

GC-MS spectrometer: An analytical tool which takes help of both Gas Chromatography and Mass Spectrometry. The combination is useful for both identification and structural elucidation of organic molecules.

Gravimetric analysis: A technique commonly used in chemical laboratories for estimation of ions by measurement of weights.

Heterogeneous catalyst: A catalyst is a substrate which makes a chemical transformation or chemical reaction faster. When the catalyst maintains a different phase from that of the reaction phase the catalyst is called a heterogeneous catalyst.

^{1}H NMR spectrometry: One type of diagnostic technique that helps structural elucidation of molecules by identifying the presence of ^{1}H-isotopes in the molecules.

HNO_3: Chemical representation of nitric acid. Nitric acid is a common laboratory reagent.

H_2SO_4: Chemical representation of sulfuric acid. Sulfuric acid is a common laboratory reagent.

K_2HPO_4: Chemical representation of potassium hydrogen phosphate. It is an important laboratory reagent used in different experiments.

Kolakhar: In Assamese language it means '*khar*' (alkali) from '*kol*' (banana plant). It is an aqueous extract of the ash prepared from banana plant. The extract is strongly alkaline. It is traditionally being used as a food additive. It is believed to have medicinal values.

Leaf sheaths: Leaf-sheaths are stacked one upon another to make the pseudo-stem of a banana plant (see Fig. **1.3**)

MMT: Million Metric Tonnes also same as million tonnes. Equivalent to 10^9 killograms.

MOP: Short form of muriate of potash. Usually it is potassium chloride mined for the use as fertilizer. Other potassium salts like potassium carbonate or potassium nitrate used as fertilizer may also be referred to as MOP.

Morphological parts of a banana plant: 'false stem' or pseudo-stem leaf-petiole, leave sheath, female flowers, hands, hanging cluster, etc. are different morphological parts of a matured and fruiting banana plant (see Figs. **1.3** and **1.4**).

MT: Metric Tonne or Tonne which is equivalent to 1000 killograms.

***Musa balbisiana* Colla** and ***Musa acuminata* Colla** These two are considered the genetically pure cultivars of banana plants. While the former bears a lot of seeds in its fruits, the latter cultivar is seedless. There are innumerable banana cultivars which are believed to be natural or artificial hybrids of the two pure cultivars.

NBPF: Native banana pseudo-stem flour which finds use both as food and medicinal ingredient.

NHB: National Horticulture Board Government of India. This board looks after the horticulture related policy and planning in India.

OHC: Oil holding capacity. It is an experimental parameter to understand the oil holding capacity of a substance.

Phenolics: Phenolics are a large group of organic molecules with phenol like properties. They are important phyto-chemicals.

Phyto-constituents: Molecular or ionic constituents found in plants.

Pith: The stem that develops at the centre of the trunk of a banana plant to push up the inflorescence is called the pith or the tender core. The pith is the stem in true sense and starts from the corm. It is rich in fiber and minerals and is used as a vegetable by many (see Fig. **1.4**).

Plantain: Please see cooking banana.

Potash: Derived from 'pot ash'. In ancient Europe people used to prepare potassium carbonate by extracting ash residues left after burning wood. Since burning was done in iron pots, the ash left in the pot was called 'pot ash', and later it became 'potash'. The word is still used to mean crude potassium carbonate mixed with other potassium salts like potassium chloride.

Potassium carbonate: It is a chemical substance which is basic in nature. In water it dissolves to afford a strongly alkaline solution. It is widely used in laboratories and industries. Natural source for potassium carbonate is virtually non-existent.

Potassium rich table salt: Table salt or the common salt available in every house-hold kitchen is called sodium chloride in chemistry. Corresponding potassium containing salt is potassium chloride. Both potassium chloride and sodium chloride are consumable. Table salt is pure sodium chloride. To make table salt rich in potassium it is mixed with potassium chloride. Potassium rich table salt is given to potassium deficient individuals or individuals suffering from hypokalemia (potassium deficiency disease).

Potassium hydroxide (KOH):	One compound of potassium; highly basic and corrosive. It absorbs carbon dioxide (CO_2) to form potassium carbonate (K_2CO_3).
Rachis:	The part of the floral stalk that is exposed between the fruits and the inflorescence with male flowers in a banana hanging cluster (see Fig. **1.3**).
SSP:	Single Super Phosphate. It is used as a chemical fertilizer to provide phosphorus to plants.
Synthetic antioxidants:	Antioxidants which are synthesized in chemical laboratories. Some other antioxidants isolated from plants are known as natural antioxidants.
TCBPF:	Tender core banana pseudo-stem flour which finds use both as food and medicinal ingredient.
TLC:	Thin Layer Chromatography. It is a low cost and fast analytical technique which can be used to identify number of molecules or ions present in a sample.
UV spectrometry:	The analytical technique that takes help of UV-radiation for diagnostic purposes.
UV-Visible spectrometry:	It is an analytical technique that takes help of UV and Visible radiations for diagnostic purposes.
WHC:	Water holding capacity. It is an experimental parameter to understand the water holding capacity of a substance.

SUBJECT INDEX

A

Absorbing moisture 123
Acetylcholine 24
Ach receptors 24
Acid 24, 26, 27, 108, 109, 129
    acetic 26
    gallic 24
    nitric 27
    oxalic 24, 129
    radicals 26, 108
Acidity 9, 23, 96
    moderate 96
Activity 24, 41
    chemo-preventive 41
    high metal chelating 24
    in vitro anti-tumor 41
Adenosine triphosphate (ATP) 51
Agricultural 45, 47, 58, 133, 134
    activities 47
    policy 45
    soils 58, 134
    use of banana plant pseudo-stem juice 133
Air 23, 25, 65, 107, 123
    dried parts 6, 107
Analgesic activity 9
Anhydride sodium sulphate 31
Anhydrous 16, 26
    diethyl ether 26
    potassium carbonate 16
Anionic radicals 33
ANOVA table 50, 102
Anti-cancer 41
    activities 41
    therapies 41
Antioxidant(s) 9, 24
    activity 9
    natural 24
    properties, multiple 24
    synthetic 24
Antivenom action 9, 134

Application of banana plant juice on farming soil 78
Aqueous extract 114, 121, 123, 136
    banana plant fiber ash 114, 123, 136
    neutralized 121
Aqueous extract 111, 112, 116, 124, 125, 127, 128
    of fiber ash 116, 125, 128
    extract of leaf sheath fiber ash 111, 116, 124, 127, 128
    of tender stem fiber ash 112, 116, 124
Aromatherapy 74
Ash 111, 114
    banana plant fibre 114
    leaf sheath fiber 111
    tender stem fiber 111
Atomic absorption 26, 31, 32, 34, 109, 111
    spectrometry 111
    spectroscopy 26, 31, 32, 34, 109
Ayurvedic system 107

B

Bacterial attack 6
Banana(s) 2, 4, 5, 10, 11, 16, 17, 23, 24, 25, 48, 106, 133
    cooking 5, 10
    crop 16, 17
    cultivars 24
    cultivation 48
    culture 11
    domesticated 5
    farmers 17
    farming 6, 17, 133
    fiber 6, 106
    fruits 10, 16, 17
    heart 2, 4
    peels 11
    stem juice 23, 24, 25
Banana leaf 5, 31
    sheath juice 31
Banana plant 2, 4, 9, 106, 107, 123, 134

www.ingramcontent.com/pod-product-compliance
Lightning Source LLC
Chambersburg PA
CBHW041709210326
41598CB00007B/591